DISEASE, SEX, COMMUNICATION, BEHAVIOR

ESSAYS IN SOCIAL BIOLOGY

Volume III

PRENTICE-HALL INTERNATIONAL, INC., *London*
PRENTICE-HALL OF AUSTRALIA, PTY. LTD., *Sydney*
PRENTICE-HALL OF CANADA, LTD., *Toronto*
PRENTICE-HALL OF INDIA PRIVATE LIMITED, *New Delhi*
PRENTICE-HALL OF JAPAN, INC., *Tokyo*

DISEASE, SEX, COMMUNICATION, BEHAVIOR

ESSAYS IN SOCIAL BIOLOGY
Volume III

Bruce Wallace

Division of Biological Sciences
Cornell University

PRENTICE-HALL, INC.
Englewood Cliffs, New Jersey

376

ISBN: 0-13-216218-0 (p)   0-13-216226-1 (c)

Library of Congress Catalogue Card Number 79-167789

Printed in the United States of America

71   18995

To MCW
DBW
RSW

Undoubtedly the most momentous piece of progress of 1969 occurred in Dijon, France, on August 14 when Neil Rappaport of Festoona, N.J., lost forever the blueprints for a machine he had just invented which would have made it possible to create a louder noise than had ever been heard before.

Acknowledgments

The author wishes to acknowledge that these essays owe their existence to a criticism made by Professor Jacques Barzun of college-level "science survey" courses. Whether I have succeeded in meeting this criticism is, of course, another matter.

A large number of colleagues at Cornell University have helped me in one way or another during the preparation of this book. Dr. Richard B. Root outlined the fascinating aspects of ecology which relate to the selection from *Iberia*. Similarly, Dr. Thomas Eisner helped me to identify many points of interest concerning communication. Dr. Richard D. O'Brien aided me in understanding the details of the transmission of nerve impulses and the modification of this transmission by chemical means. In many conversations, Dr. Gerald R. Fink explained matters related to diseases, their transmission, and the action of antibiotics. Indeed, virtually all persons who have taken part in the *Biology and Society* lectures at Cornell University have aided me immensely.

No less than my colleagues on the faculty, I must also thank the students of *Biology and Society* who undertook to read and criticize the essays during their preparation. It is a pleasure to acknowledge with special thanks the efforts of Cynthia Ravitski, Terri Schwartz, and Mark Zabek. Penny Farrow and Florence Robinson also had numerous encouraging comments and useful criticisms during this same period.

Within my own family, each member read (and seemingly enjoyed) one or more essays. Considered with my family are a host of college students who were unexpectedly pressed into reading these essays while visiting our home – both in Ithaca and abroad during the summer of 1970. Every reaction whether heeded or not has helped in the preparation of this book. And each and every one has been appreciated.

Finally, I must thank Drs. Franklin A. Long and Raymond Bowers of the Program for Science, Technology, and Society at Cornell University who arranged for a release of two months from my regular duties so that I could complete these essays. I wish to thank Dr. Harry T. Stinson for his efforts in consummating these arrangements. Financial support during these months was provided through a grant from the Alfred P. Sloan Foundation. The Program for Science, Technology, and Society has been instrumental, as well, in supporting the *Biology and Society* lectures at Cornell, a series of talks on topics similar to those discussed in my own essays.

Contents

Preface

This book is like no other biology text I know. I say this immediately to warn the reader, both my professional colleagues and the student. These essays have been written to be enjoyed. I assume that the reader has had a modern high school biology course, that he enjoys reading, and that his instructor will fill in or enlarge upon background factual material wherever necessary. Proceeding from these assumptions, I have attempted to explore contemporary biology by means of single-topic essays — touching upon a variety of questions in as informal and nontechnical a manner as possible.

From classroom experience I know that the student's initial reaction to this text is one of discomfort. Where, he asks, is the glossary of terms from which routine quizzes are prepared? What can I memorize? What am I expected to learn? What will the professor ask on his examinations? These are the conditioned reflexes of today's student — reflexes honed to a fine edge during many years of secondary and higher education. Because of the exaggerated emphasis that is placed by schools on measurable performance, much of the student's commitment to any knowledge that he might have gained in class is shed as he hands in his final exam; at that moment he has passed or failed in his effort to please his professor. Next comes the pleasure of forgetting most of the unpleasant ordeal.

Today's serious problems are not in the classroom; they are outside. These problems demand an aroused public equipped with all the information and intellectual skill that individuals can muster. We can no longer afford to shuck our knowledge at the classroom door as we depart. My plea to both the student and the instructor, therefore, is to abandon the weekly vocabulary drill and the multiple-choice examination; instead, read the essays and their accompanying selections, enjoy them, react to them, and through prolonged discussion criticize and rethink them. If enjoyment and critical discussion cause a student to retain some of what he has learned, fine. We must by all means encourage him to enjoy biology, for it is only because he may later make use of what he has retained that we teach biology to the nonbiologist.

Before World War II, bird watching and nature study made an adequate biological background for someone whose daily efforts were spent in court, in a business office, in a factory, or teaching one of the humanities in high school or college. Biology of this sort offered a pleasant diversion to the business of earning a living. Simple biology, however, is no longer sufficient. Today, many of man's problems have a biological basis — numbers of persons, the production of food to feed the hungry, man's extermination of other living things, and the wholesale destruction of the environment by man's industry. None of us can afford any longer to be a mere spectator; where a few fight for immediate profit, the many must fight for a lasting place to live. Many problems — racial, genetic, eugenic — lurking on the horizon are also biological ones. It is ridiculous to expect intelligent action to come from biologically naive persons who, with fatal optimism, seek at most ad hoc solutions to grave problems. The decisions affecting irreversible governmental and industrial actions are largely scientific in nature; we can no longer be satisfied with science survey courses that merely train dillettante bird watchers and butterfly collectors. These courses must teach students to grapple with the serious problems of today — and tomorrow.

What precisely is the aim of a required course in elementary biology at the college level? What is it that the student should learn? Opinions differ. I believe that today's college student most likely has had a modern course in high school biology comparable to those prepared by the Biological Sciences Curriculum Study. I feel that he should understand the workings of his body and mind ("know thyself" is still an apt phrase) and how both continue to work under a barrage of external and internal challenges. He should see how mankind fits into the skein of life on earth and appreciate the constraints placed upon man if both he and the skein are to continue their existence. We are, as the astronauts said, confined to a raft that is a small blue gem floating in space; our problem is to keep the raft gemlike and blue. Finally, the student should develop a sense of understanding and a method of intellectual attack appropriate for the problems that confront (haunt?) us in our daily relationship with government, business, community, and neighborhood affairs. The continual reliance upon expedient solutions to grave problems resembles the instant-by-instant reflexive responses of a frightened hare attempting to elude a dog; it does not become the intellectual powers of man. A college-level biology course for nonmajors should help the individual visualize and evaluate probable outcomes of certain courses of action. It should teach him to cope with and refute the oldest and most seductive of all ad hoc arguments: "It smells like money."

How does one obtain these goals? In truth, no one knows. To ask each student to make personal observations on all matters of import is patently foolish; in our complex society such observations must be made vicariously. Consequently, in the sections that follow I have drawn upon the observations and experiences made and described not by scientists but by essayists, novelists, and historians. I have turned to these persons because they are not only professional observers but also are competent writers. In doing this I may have risked scientific precision but I have gained tremendously in other respects. A clinical account, no matter how accurate, cannot match Voltaire when he writes "a beggar all covered with sores, dead-eyed, the end of his nose eaten away, his mouth deformed, his teeth black, tormented with a violent cough . . . ."

Because students are initially discomforted by my approach, I want to explain further my use of literary selections in a science text. In my view they substitute for personal observations but not for the precise and structured observations of a field trip. They more nearly resemble series of strolls, some brief and others perhaps overly long. Each establishes a mood and offers an excuse to embark upon a series of discussions; the essays serve this latter purpose. No attempt has been made to write an essay about every possible topic posed by its accompanying selection nor have the essays been restricted to topics included within the selection. If the selections can be compared with strolls, the essays can be compared with the conversational ramblings of a talkative companion — one inspired by events he encounters but whose conversation is neither restricted to nor necessarily evoked by the passing scene.

The reader will find in these three volumes selections from a number of sources as well as a variety of single-topic essays. Their arrangement, of necessity, is not rigorously sequential. Sometimes in earlier essays I have omitted details needed to understand later ones. Sometimes I have repeated information. In any case, the effect is to make the essays just the more chatty and repetitious. When the reader has finished the last essay, he will be totally uninformed of such things as the difference between *polyploidy, polyteny,* and *polysomy,* but why shouldn't he be? There is a remote chance that he will have enjoyed himself at some point. If his enjoyment causes him to retain a knack for wrestling with today's problems, and if he uses this knack by speaking up publicly on controversial environmental and population issues, his enjoyment could prove to be the lasting contribution of the book.

Bruce Wallace

A Word to the Teacher

The use of this text will call for techniques and attitudes unlike those found in most biology classes. Student interest in the types of topics discussed in these essays is nearly overwhelming. Lecturers in the initial *Biology and Society* series at Cornell University (Biology 201-202, 2 points per semester) found themselves addressing sustained weekly audiences of 1,000 (students and townspeople) rather than the 25 or 30 that most had expected. The problems that arise from an enthusiastic response of this kind are of three sorts: (1) the handling of large numbers of students in an intimate way; (2) the choice of factual material to be stressed in the course; (3) the basis of grading students who enroll for credit.

How can large numbers of students be handled where the burden of a social biology course falls on one staff member? I suggest that this man (presumably a biologist) be conversant with the material covered by any one of a number of excellent college-level biology textbooks and, further, with the additional material to be found in some 130 *Scientific American* offprints. Each lecture should be restricted to no more than 20 minutes; it should present only the facts that are essential for understanding the assigned essay(s). The remaining 30 minutes should be reserved for student discussion and questions from the floor. The role of the lecturer during this period is to encourage student participation and to supply factual information when necessary in order to keep the discussion accurate. Where information is lacking or when someone questions the accuracy of an essay, a team of two or three students should be told to recheck the facts and return in a week's time with a five- or ten-minute report for the class.

Facts and the collection of factual material should be recurrent themes throughout all discussions. The types of facts, however, that are important in a type of course like *Biology and Society* need not be those that are considered important in most beginning biology courses. Vocabulary is trivial, except to the biology major whose livelihood will depend upon knowing it. Many of the details of ordinary biology courses are also trivial – especially those details that are rapidly forgotten by most students and, consequently, are so useful as test items. The facts that I feel are important are the obvious ones that tend to be neglected. These are the facts that at times are embarrassing to teach because the students react by saying, "Of course." They must be repeated again and again nevertheless. If simple facts were understood by everyone, permanent equipment would not be designed to wear out nor would disposable articles be made so often of indestructible materials.

My personal feelings on the matter of grades and grading are rather strong ones. The student who attends a social biology course and later, because of it, takes an active and intelligent part in community affairs attests to the success of the course and its teacher; the student who memorizes facts, talks in class, but refuses to become involved in society's affairs represents a failure. I have argued that grades for Cornell's *Biology and Society* should be limited to S/U (satisfactory-unsatisfactory) *only* and that those who take the course should receive credit only for use in accumulating general credits. This argument is based in part on my feeling that the choice of lecture topics and the overall format of the course can most likely remain flexible and stimulating only if traditional administrative groups and committees are kept uninvolved; once a standing committee feels that the course content falls within its domain, freedom to arrange novel discussions or sequences of off-beat lecture topics may disappear. In addition, however, I firmly believe that no appropriate grade in social biology can be given until at least two decades after the student's graduation.

SECTION ONE

Disease

Introduction

Death and taxes! Man's constant companions. No one talks seriously about abolishing taxes; occasionally someone does speak of outlawing death. What a villain he would be if he were to succeed — despised and cursed for his folly. To abolish death before abolishing birth would be calamitous; to abolish death before abolishing senility would be a travesty; to abolish death before abolishing boredom would be torture.

The selections chosen for the section of disease have come from a variety of sources and have been included for a variety of reasons. *The Journal of the Plague Year* has been called a tale of horror told by one of the great masters of realism; in its pages a metropolis dies before our eyes. The irresistible progress of the plague as it swept across London — across a single city — is painstakingly recorded in the excerpts reprinted here. Recorded, too, are the small signs that foretold the end of the epidemic after an estimated 100,000 corpses had been dumped into common graves.

Infectious diseases, those that travel from one person to another, are not independent agents; man can do very well without them, but they need man desperately. This dependence of disease organisms on man underlies a series of interlocking evolutionary modifications on the part of both the disease germ and its host: the germ changes its effects so that it becomes tolerable; man increases his resistance so that the effects of the disease are minimized. The selection from *Hawaii* describes an outbreak of measles in an exotic Polynesian population. What is an innocuous childhood disease for the recently arrived white missionary and his family causes the resident doctor to cry out in alarm.

The selection from *My Life as a Young Girl* mentions almost by accident a fatal case of tetanus; the death of a neighbor's child is innocently included as a fragment of a single day's entry in a young girl's diary of nearly a century ago. Tetanus, anthrax, and gangrene as well as some other serious infections differ from the plague or measles because, unlike the latter, they are not spread from person to person. If a disease is spread by contagion, the person who moves about even though he is ill is the one who infects the greatest number of others; consequently, the organism that spreads most rapidly through a population is the one that affects its victims the least. This simple relationship explains why the effects of an infectious disease become less devastating over a period of time. Anthrax, tetanus, and gangrene are diseases, however, that arise following an infection by long-lived spores of the soil. All three are serious, often fatal, diseases. The responsible organisms are under no obligation to mankind. They

3

infect him only by accident. Moreover, their spread, their very existence, is independent of the ultimate fate — life or death — of infected human beings.

For a transition from infectious to degenerative diseases, from diseases caused by microorganisms to those caused by the breakdown of bodily organization and function, we shall look at the lengthy selection reproduced from *The Death of Ivan Ilych.* Ivan Ilych died of abdominal cancer; there is no mistaking Tolstoy's description. Diagnostic procedures are immensely better today than they were in the 19th century, and so cancer can now be detected earlier and treated more effectively. Nevertheless, the progress of untreated cancer today is as relentless and terrifying as it was during Tolstoy's life. In *The Death of Ivan Ilych,* Tolstoy has set down in stark detail an account of what passes as a "long illness" in the discreet obituary.

Ivan Ilych was snatched from life by an ailment he did not understand. He was eased out of existence by an impatient family and embarrassed friends who could not admit, until death was imminent, that he was indeed dying. Tolstoy's account of Ivan's death has been used here as a pretext for discussing degenerative diseases as an increasingly important cause of death for man. In his essay "On Being Finite" in this volume, J.B.S. Haldane points out that many wild animals scarcely ever die of old age because accidental death almost always intervenes.

What is the role of death in nature? Why are individual members of all species mortal? Or are they? And, how should the inevitability of death affect our attitude toward it? These questions are raised in the two concluding essays of this section. As a selection to accompany these concluding essays, perhaps as a counterfoil to my own essay "On the Dignity of Dying," I have included Hemingway's "A Natural History of the Dead" from *Death in the Afternoon.*

A Journal of
the Plague Year

Daniel Defoe

It was about the beginning of September, 1664, that I, among the rest of my neighbours, heard, in ordinary discourse, that the plague was returned again in Holland; for it had been very violent there, and particularly at Amsterdam and Rotterdam, in the year 1663, whither, they say, it was brought, some said from Italy, others from the Levant, among some goods, which were brought home by their Turkey fleet; others said it was brought from Candia; others from Cyprus. It mattered not from whence it came; but all agreed it was come into Holland again.

We had no such thing as printed newspapers in those days to spread rumours and reports of things, and to improve them by the invention of men, as I have lived to see practised since. But such things as those were gathered from the letters of merchants and others who corresponded abroad, and from them was handed about by word of mouth only; so that things did not spread instantly over the whole nation, as they do now. But it seems that the Government had a true account of it, and several councils were held about ways to prevent its coming over; but all was kept very private. Hence it was that this rumour died off again, and people began to forget it, as a thing we were very little concerned in, and that we hoped was not true, till the latter end of November or the beginning of December, 1664, when two men, said to be Frenchmen, died of the plague in Long Acre, or rather at the upper end of Drury Lane. The family they were in endeavoured to conceal it as much as possible; but as it had gotten some vent in the discourse of the neighbourhood, the Secretaries of State got knowledge of it, and concerning themselves to enquire about it, in order to be certain of the truth, two physicians and a surgeon were ordered to go to the house and make inspection. This they did; and finding evident tokens of the sickness upon both the bodies that were dead, they gave their opinions publickly that they died of the plague. Whereupon it was given in to the parish clerk, and he also returned them to the Hall; and it was printed in the weekly bill of mortality in the usual manner, thus —

Plague, 2. Parishes infected, 1.

The people shewed a great concern at this, and began to be alarmed all over the town, and the more, because in the last week in December, 1664, another man died in the same house, and of the same distemper. And then we were easy again for about six weeks, when none having died with any marks of infection, it was said the distemper was gone; but after that, I think it was about the 12th of February, another died in another house, but in the same parish and in the same manner.

This turned the people's eyes pretty much towards that end of the town; and the weekly bills shewing an increase of burials in St Giles's parish more than usual, it began to be suspected that the plague was among the people at that end of the town, and that many had died of it, though they had taken care to keep it as much from the knowledge of the publick as possible. This possessed the heads of the people very much, and few cared to go through Drury Lane, or the other streets suspected, unless they had extraordinary business that obliged them to it.

This increase of the bills stood thus: the usual number of burials in a week, in the parishes of St Giles-in-the-Fields and St Andrew Holborn, were from twelve to seventeen or nineteen each, few more or less; but from the time that the plague first began in St Giles's parish, it was observed that the ordinary burials increased in number considerably. For example:

From December 27 to January	3	.	St Giles's . .	16
			St Andrew's .	17
" January 3 " "	10	.	St Giles's . .	12
			St Andrew's .	25
" January 10 " "	17	.	St Giles's . .	18
			St Andrew's .	18
" January 17 " "	24	.	St Giles's . .	23
			St Andrew's .	16
" January 24 " "	31	.	St Giles's . .	24
			St Andrew's .	15
" January 30 " February	7	.	St Giles's . .	21
			St Andrew's .	23
" February 7 " "	14	.	St Giles's . .	24

Whereof one of the plague.

The like increase of the bills was observed in the parishes of St Bride, adjoining on one side of Holborn parish, and in the parish of St James Clerkenwell, adjoining on the other side of Holborn; in both which parishes the usual numbers that died weekly were from four to six or eight, whereas at that time they were increased as follows:

From December 20 to December 27 .	{ St Bride's . .	0
	St James's . .	8
" December 27 to January 3 .	{ St Bride's . .	6
	St James's . .	9
" January 3 " " 10 .	{ St Bride's . .	11
	St James's . .	7
" January 10 " " 17 .	{ St Bride's . .	12
	St James's . .	9
" January 17 " " 24 .	{ St Bride's . .	9
	St James's . .	15
" January 24 " " 31 .	{ St Bride's . .	8
	St Jame's . .	12
" January 31 " February 7 .	{ St Bride's . .	13
	St James's . .	5
" February 7 " " 14 .	{ St Bride's . .	12
	St James's . .	6

Besides this, it was observed with great uneasiness by the people that the weekly bills in general increased very much during these weeks, although it was at a time of the year when usually the bills are very moderate.

The usual number of burials within the bills of mortality for a week was from about 240 or thereabouts to 300. The last was esteemed a pretty high bill; but after this we found the bills successively increasing, as follows:

	Buried.	Increased.
December the 20th to the 27th	291	. . . .
" 27th " 3rd January	. . 349	. . 58
January the 3rd " 10th "	. . 394	. . 45
" 10th " 17th "	. . 415	. . 21
" 17th " 24th "	. . 474	. . 59

This last bill was really frightful, being a higher number than had been known to have been buried in one week since the preceding visitation of 1656.

However, all this went off again, and the weather proving cold, and the frost, which began in December, still continuing very severe, even till near the end of February, attended with sharp though moderate winds, the bills decreased again, and the city grew healthy, and everybody began to look upon the danger as good as over; only that still the burials in St Giles's continued high. From the beginning of April especially they stood at twenty-five each week, till the week from the 18th to the 25th, when there was buried in St Giles's parish thirty, whereof two of the plague and eight of the spotted fever, which was looked upon as the same thing; likewise the number that died of the spotted fever in the whole increased, being eight the week before, and twelve the week above named.

This alarmed us all again, and terrible apprehensions were among the people, especially the weather being now changed and growing warm, and the summer being at hand. However, the next week there seemed to be some hopes again; the bills were low, the number of the dead in all was but 388, there was none of the plague, and but four of the spotted fever.

But the following week it returned again, and the distemper was spread into two or three other parishes, viz., St Andrew's Holborn; St Clement Danes; and, to the great affliction of the city, one died within the walls, in the parish of St Mary Woolchurch, that is to say, in Bearbinder Lane, near Stocks Market; in all there were nine of the plague and six of the spotted fever. It was, however, upon enquiry, found that this Frenchman who died in Bearbinder Lane was one who, having lived in Long Acre, near the infected houses, had removed for fear of the distemper, not knowing that he was already infected.

This was the beginning of May, yet the weather was temperate, variable, and cool enough, and people had still some hopes. That which encouraged them was that the city was healthy, the whole ninety-seven parishes buried but fifty-four, and we began to hope, that as it was chiefly among the people at that end of the town, it might go no farther; and the rather, because the next week, which was from the 9th of May to the 16th, there died but three, of which not one within the whole city or liberties; and St Andrew's buried but fifteen, which was very low. 'Tis true St Giles's buried two-and-thirty, but still, as there was but one of the plague, people began to be easy. The whole bill also was very low, for the week before the bill was but 347, and the week above mentioned but 343. We continued in these hopes for a few days. But it was but for a few, for the people were no more to be deceived thus; they searched the houses, and found that the plague was really spread every way, and that many died of it every day. So that now all our extenuations abated, and it was no more to be concealed; nay, it quickly appeared that the infection had spread itself beyond all hopes of abatement; that in the parish of St Giles it was gotten into several streets, and several families lay all sick together. And, accordingly, in the weekly bill for the next week the thing began to shew itself; there was indeed but fourteen set down of the plague, but this was all knavery and collusion, for [in] St Giles's parish they buried forty in all, whereof it was certain most of them died of the plague, though they were set down of other distempers; and though the number of all the burials were not increased above thirty-two, and the whole bill being but 385, yet there was fourteen of the spotted fever, as well as fourteen of the plague; and we took it for granted upon the whole that there were fifty died that week of the plague.

The next bill was from the 23rd of May to the 30th, when the number of the plague was seventeen. But the burials in St Giles's were fifty-three — a frightful number! — of whom they set down but nine of the plague. But on an examination more strictly by the justices of the peace, and at the Lord Mayor's request, it was found there were twenty more who were really dead of the plague in that parish, but had been set down of the spotted fever or other distempers, besides others concealed.

But those were trifling things to what followed immediately after; for now the weather set in hot, and from the first week in June the infection spread in a dreadful manner, and the bills rose high; the articles of the fever, spotted fever, and teeth began to swell; for all that could conceal their distempers did it, to prevent their neighbours shunning and refusing to converse with them, and also to prevent authority shutting up their houses, which though it was not yet practised, yet was threatened, and people were extremely terrified at the thoughts of it.

The second week in June, the parish of St Giles, where still the weight of the infection lay, buried 120, whereof, though the bills said but 68 of the plague, everybody said there had been 100 at least, calculating it from the usual number of funerals in that parish, as above.

Till this week the city continued free, there having never any died, except that one Frenchman who I mentioned before, within the whole ninety-seven parishes. Now there died four within the city, one in Wood Street, one in Fenchurch Street, and two in Crooked Lane. Southwark was entirely free, having not one yet died on that side of the water.

I lived without Aldgate, about midway between Aldgate Church and Whitechapel Bars, on the left hand or north side of the street; and as the distemper had not reached to that side of the city, our neighbourhood continued very easy. But at the other end of the town their consternation was very great; and the richer sort of people, especially the nobility and gentry from the west part of the city, thronged out of town with their families and servants in an unusual manner; and this was more particularly seen in Whitechapel; that is to say, the broad street where I lived; indeed, nothing was to be seen but wagons and carts, with goods, women, servants, children, &c.; coaches filled with people of the better sort, and horsemen attending them, and all hurrying away; then empty waggons and carts appeared, and spare horses with servants, who, it was apparent, were returning or sent from the countries to fetch more people; besides innumerable numbers of men on horseback, some alone, others with servants, and, generally speaking, all loaded with baggage and fitted out for travelling, as any one might perceive by their appearance.

This was a very terrible and melancholy thing to see, and as it was a sight which I could not but look on from morning to night, for indeed there was nothing else of moment to be seen, it filled me with very serious thoughts of the misery that was coming upon the city, and the unhappy condition of those that would be left in it.

This hurry of the people was such for some weeks that there was no getting at the Lord Mayor's door without exceeding difficulty; there was such pressing and crowding there to get passes and certificates of health for such as travelled abroad, for without these there was no being admitted to pass through the towns upon the road, or to lodge in any inn. Now, as there had none died in the city for all this time, my Lord Mayor gave certificates of health without any difficulty to all those who lived in the ninety-seven parishes, and to those within the liberties too for a while.

This hurry, I say, continued some weeks, that is to say, all the month of May and June, and the more because it was rumoured that an order of the Government was to be issued out to place turnpikes and barriers on the road to prevent people's travelling, and that the towns on the road would not suffer people from London to pass for fear of bringing the infection along with them, though neither of these rumours had any foundation but in the imagination, especially at first.

• • •

It was now mid-July, and the plague, which had chiefly raged at the other end of the town, and, as I said before, in the parishes of St Giles, St Andrew Holborn, and towards Westminster, began to now come eastward towards the part where I lived. It was to be observed, indeed, that it did not come straight on towards us; for the city, that is to say, within the walls, was indifferently healthy still; nor was it got then very much over the water into Southwark; for though, there died that week 1,268 of all distempers, whereof it might be supposed above 900 died of the plague, yet there was but 28 in the whole city, within the walls, and but 19 in Southwark, Lambeth parish included; whereas in the parishes of St Giles and St Martin-in-the-Fields alone there died 421.

But we perceived the infection chiefly in the outparishes, which being very populous, and fuller also of poor, the distemper found more to prey upon than in the city, as I shall observe afterwards. We perceived, I say, the distemper to draw our way, viz., by the parishes of Clerkenwell, Cripplegate, Shoreditch, and Bishopsgate; which last two parishes joining to Aldgate, Whitechapel, and Stepney, the infection came at length to spread its utmost rage and violence in those parts, even when it abated at the western parishes where it began.

It was very strange to observe that in this particular week, from the 4th to the 11th of July, when, as I have observed, there died near four hundred of the plague in the two parishes of St Martin and St Giles-in-the-Fields only, there died in the parish of Aldgate but four, in the parish of Whitechapel three, in the parish of Stepney but one.

Likewise in the next week, from the 11th of July to the 18th, when the week's bill was 1,761, yet there died no more of the plague, on the whole Southwark side of the water, than sixteen.

But this face of things soon changed, and it began to thicken in Cripplegate parish especially, and in Clerkenwell; so that by the second week in August, Cripplegate parish alone buried 886, and Clerkenwell 155. Of the first,

850 might well be reckoned to die of the plague; and of the last, the bill itself said 145 were of the plague.

During the month of July, and while, as I have observed, our part of the town seemed to be spared in comparison of the west part, I went ordinarily about the streets, as my business required, and particularly went generally once in a day, or in two days, into the city, to my brother's house, which he had given me charge of, and to see if it was safe. And having the key in my pocket, I used to go into the house, and over most of the rooms, to see that all was well; for though it be something wonderful to tell, that any should have hearts so hardened in the midst of such a calamity as to rob and steal, yet certain it is that all sorts of villainies, and even levities and debaucheries, were then practised in the town as openly as ever — I will not say quite as frequently, because the numbers of people were many ways lessened.

But the city itself began now to be visited too, I mean within the walls; but the number of people there were indeed extremely lessened by so great a multitude having been gone into the country; and even all this month of July they continued to flee, though not in such multitudes as formerly. In August, indeed, they fled in such a manner that I began to think there would be really none but magistrates and servants left in the city.

As they fled now out of the city, so I should observe that the Court removed early, viz., in the month of June, and went to Oxford, where it pleased God to preserve them; and the distemper did not, as I heard of, so much as touch them, for which I cannot say that I every saw they shewed any great token of thankfulness, and hardly anything of reformation, though they did not want being told that their crying vices might, without breach of charity, be said to have gone far in bringing that terrible judgment upon the whole nation.

The face of London was now indeed strangely altered, I mean the whole mass of buildings, city, liberties, suburbs, Westminster, Southwark, and altogether; for as to the particular part called the city, or within the walls, that was not yet much infected. But in the whole the face of things, I say, was much altered; sorrow and sadness sat upon every face; and though some part were not yet overwhelmed, yet all looked deeply concerned; and as we saw it apparently coming on, so every one looked on himself and his family as in the utmost danger: were it possible to represent those times exactly to those that did not see them, and give the reader due ideas of the horror that everywhere presented itself, it must make just impressions upon their minds and fill them with surprize. London might well be said to be all in tears; the mourners did not go about the streets indeed, for nobody put on black or made a formal dress of mourning for their nearest friends; but the voice of mourning was truly heard in the streets; the shrieks of women and children at the windows and doors of their houses, where their dearest relations were perhaps dying, or just dead, were so frequent to be heard as we passed the streets, that it was enough to pierce the stoutest heart in the world to hear them. Tears and lamentations were seen almost in every house, especially in the first part of the visitation; for towards the latter end men's hearts were hardened, and death was so always before their eyes, that they did not so much concern themselves for the loss of their friends, expecting that themselves should be summoned the next hour.

Business led me out sometimes to the other end of the town, even when the sickness was chiefly there; and as the thing was new to me, as well as to everybody else, it was a most surprizing thing to see those streets which were usually so thronged now grown desolate, and so few people to be seen in them, that if I had been a stranger and at a loss for my way, I might sometimes have gone the length of a whole street, I mean of the by-streets, and seen nobody to direct me except watchmen set at the doors of such houses as were shut up, of which I shall speak presently.

One day, being at that part of the town on some special business, curiosity led me to observe things more than usually, and indeed I walked a great way where I had no business; I went up Holborn, and there the street was full of people, but they walked in the middle of the great street, neither on one side or other, because, as I suppose, they would not mingle with anybody that came out of houses, or meet with smells and scents from houses that might be infected.

The Inns of Court were all shut up; nor were very many of the lawyers in the Temple, or Lincoln's Inn, or Grey's Inn, to be seen there. Everybody was at peace; there was no occasion for lawyers; besides, it being in the time of the vacation too, they were generally gone into the country. Whole rows of houses in some places were shut close up, the inhabitants all fled, and only a watchman or two left.

When I speak of rows of houses being shut up, I do not mean shut up by the magistrates, but that great numbers of persons followed the Court, by the necessity of their employments and other dependencies; and as others retired, really frighted with the distemper, it was a meer desolating of some of the streets. But the fright was not yet near so great in the city, abstractly so called, and particularly because, though they were at first in a most inexpressible consternation, yet as I have observed that the distemper intermitted often at first, so they were, as it were, alarmed and unalarmed again, and this several times, till it began to be familiar to them; and that even when it appeared violent, yet seeing it did not presently spread into the city, or the east and south parts, the people began to take courage, and to be, as I may say, a little hardened. It is true a vast many people fled, as I have observed, yet they were chiefly from the west end of the town, and from that we call the heart of the city, that is to say, among the wealthiest of the people, and such people as were unencumbered with trades and business. But of the rest, the generality stayed, and seemed to abide the worst; so that in the place we call the liberties, and in the suburbs, in Southwark, and in the east part, such as Wapping, Ratcliff, Stepney, Rotherhithe, and the like, the people generally stayed, except here and there a few wealthy families, who, as above, did not depend upon their business.

●   ●   ●

All the plays and interludes which, after the manner of the French Court, had been set up and began to increase among us, were forbid to act; the gaming tables, publick dancing rooms, and music houses, which multiplied and began to debauch the manners of the people, were shut up and suppressed; and the jack-puddings, merry-andrews, puppet-shows, rope-dancers, and such-like doings, which had bewitched the poor common people, shut up their shops, finding indeed no trade; for the minds of the people were agitated with other things, and a kind of sadness and horror at these things sat upon the countenances even of the common people. Death was before their eyes, and everybody began to think of their graves, not of mirth and diversions.

But even those wholesome reflections, which, rightly managed, would have most happily led the people to fall upon their knees, make confession of their sins, and look up to their merciful Saviour for pardon, imploring His compassion on them in such a time of their distress, by which we might have been as a second Nineveh, had a quite contrary extream in the common people, who, ignorant and stupid in their reflections as they were brutishly wicked and thoughtless before, were now led by their fright to extreams of folly; and, as I have said before that they ran to conjurers and witches, and all sorts of deceivers, to know what should become of them (who fed their fears, and kept them always alarmed and awake on purpose to delude them and pick their pockets), so they were as mad upon their running after quacks and mountebanks and every practising old woman for medicines and remedies, storing themselves with such multitudes of pills, potions, and preservatives, as they were called, that they not only spent their money, but even poisoned themselves beforehand for fear of the poison of the infection, and prepared their bodies for the plague instead of preserving them against it. On the other hand, it is incredible, and scarce to be imagined, how the posts of houses and corners of streets were plastered over with doctors' bills and papers of ignorant fellows, quacking and tampering in physick, and inviting the people to come to them for remedies, which was generally set off with such flourishes as these, viz.: "INFALLIBLE preventive pills against the plague." "NEVER-FAILING preservatives against the infection." "SOVEREIGN cordials against the corruption of the air." "EXACT regulations for the conduct of the body in case of an infection." "Anti-pestilential pills." "INCOMPARABLE drink against the plague, never found out before." "An UNIVERSAL remedy for the plague." "The ONLY TRUE plague-water." "The ROYAL ANTIDOTE against all kinds of infection;" and such a number more that I cannot reckon up; and if I could, would fill a book of themselves to set them down.

Others set up bills to summon people to their lodgings for directions and advice in the case of infection. These had specious titles, such as these:

> An eminent High Dutch physician, newly come over from Holland, where he resided during all the time of the great plague last year in Amsterdam, and cured multitudes of people that actually had the plague upon them.

An Italian gentlewoman just arrived from Naples, having a choice secret to prevent infection, which she found out by her great experience, and did wonderful cures with it in the late plague there, wherein there died 20,000 in one day.

An antient gentlewoman, having practised with great success in the late plague in this city, anno 1636, gives her advice only to the female sex. To be spoke with, &c.

An experienced physician, who has long studied the doctrine of antidotes against all sorts of poison and infection, has, after forty years' practice, arrived to such skill as may, with God's blessing, direct persons how to prevent their being touched by any contagious distemper whatsoever. He directs the poor gratis.

I take notice of these by way of specimen. I could give you two or three dozen of the like and yet have abundance left behind. 'Tis sufficient from these to apprise any one of the humour of those times, and how a set of thieves and pickpockets not only robbed and cheated the poor people of their money, but poisoned their bodies with odious and fatal preparations; some with mercury, and some with other things as bad, perfectly remote from the thing pretended to, and rather hurtful than serviceable to the body in case an infection followed.

I cannot omit a subtility of one of those quack operators, with which he gulled the poor people to crowd about him, but did nothing for them without money. He had, it seems, added to his bills, which he gave about the streets, this advertisement in capital letters, viz., "He gives advice to the poor for nothing."

Abundance of poor people came to him accordingly, to whom he made a great many fine speeches, examined them of the state of their health and of the constitution of their bodies, and told them many good things for them to do, which were of no great moment. But the issue and conclusion of all was, that he had a preparation which if they took such a quantity of every morning, he would pawn his life they should never have the plague; no, though they lived in the house with people that were infected. This made the people all resolve to have it; but then the price of that was so much, I think 'twas half-a-crown. "But, sir," says one poor woman, "I am a poor almswoman, and am kept by the parish, and your bills say you give the poor your help for nothing." "Ay, good woman," says the doctor, "so I do, as I published there. I give my advice to the poor for nothing, but not my physick." "Alas, sir!" says she, "that is a snare laid for the poor, then; for you give them advice for nothing; that is to say, you advise them gratis, to buy your physick for their money; so does every shopkeeper with his wares." Here the woman began to give him ill words, and stood at his door all that day, telling her tale to all the people that came, till the doctor, finding she turned away his customers, was obliged to call her upstairs again and give her his box of physick for nothing, which perhaps, too, was good for nothing when she had it.

But to return to the people, whose confusions fitted them to be imposed upon by all sorts of pretenders and by every mountebank. There is no doubt but these quacking sort of fellows raised great gains out of the miserable people, for we daily found the crowds that ran after them were infinitely greater, and their doors were more thronged than those of Dr Brooks, Dr Upton, Dr Hodges, Dr Berwick, or any, though the most famous men of the time. And I was told that some of them got five pounds a day by their physick.

But there was still another madness beyond all this, which may serve to give an idea of the distracted humour of the poor people at that time, and this was their following a worse sort of deceivers than any of these; for these petty thieves only deluded them to pick their pockets and get their money, in which their wickedness, whatever it was, lay chiefly on the side of the deceivers deceiving, not upon the deceived. But in this part I am going to mention it lay chiefly in the people deceived, or equally in both, and this was in wearing charms, philtres, exorcisms, amulets, and I know not what preparations, to fortify the body with them against the plague; as if the plague was not the hand of God, but a kind of a possession of an evil spirit, and that it was to be kept off with crossings, signs of the zodiac, papers tied up with so many knots, and certain words or figures written on them, as particularly the word Abracadabra, formed in triangle or pyramid, thus:

ABRACADABRA	
ABRACADABR	Others had the Jesuits'
ABRACADAB	mark in a cross:
ABRACADA	I H
ABRACAD	S.
ABRACA	
ABRAC	Others nothing but this
ABRA	mark, thus:
ABR	
AB	
A	

I might spend a great deal of time in my exclamations against the follies, and indeed the wickedness, of those things in a time of such danger, in a matter of such consequences as this, of a national infection. But my memorandums of these things relate rather to take notice only of the fact, and mention only that it was so. How the poor people found the insufficiency of those things, and how many of them were afterwards carried away in the dead-carts and thrown into the common graves of every parish with these hellish charms and trumpery hanging about their necks, remains to be spoken of as we go along.

•   •   •

It was observable, then, that this calamity of the people made them very humble; for now for about nine weeks together there died near a thousand a day, one day with another, even by the account of the weekly bills, which yet, I have reason to be assured, never gave a full account, by many thousands, the confusion being such, and the carts working in the dark when they carried the dead, that in some places no account at all was kept, but they worked on, the clerks and sextons not attending for weeks together, and not knowing what number they carried. This account is verified by the following bills of mortality:

	Of all Diseases.	Of the Plague.
From August 8 to August 15	5,319	3,880
” ” 15 ” 22	5,568	4,237
” ” 22 ” 29	7,496	6,102
” ” 29 to September 5	8,252	6,988
” Sepember 5 ” 12	7,690	6,544
” ” 12 ” 19	8,297	7,165
” ” 19 ” 26	6,460	5,533
” ” 26 to October 3	5,720	4,929
” October 3 ” 10	5,068	4,327
	59,870	49,705

So that the gross of the people were carried off in these two months; for, as the whole number which was brought in to die of the plague was but 68,590, here is 50,000 of them, within a trifle, in two months; I say 50,000, because, as there wants 295 in the number above, so there wants two days [sic] of two months in the account of time.

Now, when I say that the parish officers did not give in a full account, or were not to be depended upon for their account, let any one but consider how men could be exact in such a time of dreadful distress, and when many of them were taken sick themselves, and perhaps died in the very time when their accounts were to be given in; I mean the parish clerks, besides inferior officers; for though these poor men ventured at all hazards, yet they were far from being exempt from the common calamity, especially if it be true that the parish of Stepney had, within the year, 116 sextons, gravediggers, and their assistants; that is to say, bearers, bellmen, and drivers of carts for carrying off the dead bodies.

Indeed the work was not of a nature to allow them leisure to take an exact tale of the dead bodies, which were all huddled together in the dark into a pit; which pit or trench no man could come nigh but at the utmost peril. I observed often that in the parishes of Aldgate and Cripplegate, Whitechapel and Stepney, there were five, six, seven, and eight hundred in a week in the bills, whereas, if we may believe the opinion of those that lived in the city all the time as well as I, there died sometimes two thousand a week in those parishes; and I saw it under the hand of one that made as strict an examination into that part as he could, that there really died an hundred thousand people of the plague in that one year, whereas in the bills, the articles of the plague, it was but 68,590.

If I may be allowed to give my opinion, by what I saw with my eyes and heard from other people that were eye-witnesses, I do verily believe the same, viz., that there died at least an hundred thousand of the plague only, besides other distempers, and besides those which died in the fields and highways and secret places out of the compass of the communication, as it was called, and who were not put down in the bills, though they really belong to the body of the inhabitants. It was known to us all that abundance of poor despairing creatures who had the distemper upon them, and were grown stupid or melancholy by their misery, as many were, wandered away into the fields and woods, and into secret, uncouth places, almost anywhere, to creep into a bush or hedge and die.

•　•　•

It remains now that I should say something of the merciful part of this terrible judgement. The last week in September, the plague being come to its crisis, its fury began to assuage. I remember my friend, Dr Heath, coming to see me the week before, told me he was sure that the violence of it would assuage in a few days; but when I saw the weekly bill of that week, which was the highest of the whole year, being 8,297 of all diseases, I upbraided him with it, and asked him what he had made his judgement from. His answer, however, was not so much to seek as I thought it would have been. "Look you," says he, "by the number which are at this time sick and infected, there should have been twenty thousand dead the last week instead of eight thousand, if the inveterate mortal contagion had been as it was two weeks ago; for then it ordinarily killed in two or three days, now not under eight or ten; and then not above one in five recovered, whereas I have observed that now not above two in five miscarry. And, observe it from me, the next bill will decrease, and you will see many more people recover than used to do; for though a vast multitude are now everywhere infected, and as many every day fall sick, yet there will not so many die as there did, for the malignity of the distemper is abated;" adding that he began now to hope, nay, more than hope, that the infection had passed its crisis and was going off; and accordingly so it was, for the next week being, as I said, the last in September, the bill decreased almost two thousand.

It is true the plague was still at a frightful height, and the next bill was no less than 6,460 and the next to that, 5,720; but still my friend's observation was just, and it did appear the people did recover faster and more in number than they used to do; and indeed, if it had not been so, what had been the condition of the city of London? For, according to my friend, there were not fewer than 60,000 people at that time infected, whereof, as above 20,477 died, and near 40,000 recovered; whereas, had it been as it was before 50,000 of that number would very probably have died, if not more, and 50,000 more would have sickened; for, in a word, the whole mass of people began to sicken, and it looked as if none would escape.

But this remark of my friend's appeared more evident in a few weeks more, for the decrease went on, and another week in October it decreased 1,843, so that the number dead of the plague was but 2,665; and the next week it decreased 1,413 more, and yet it was seen plainly that there was abundance of people sick, nay, abundance more than ordinary, and abundance fell sick every day but (as above) the malignity of the disease abated.

Such is the precipitant disposition of our people . . . , that as upon the first fright of the infection, they shunned one another, and fled from one another's houses and from the city with an unaccountable and, as I thought, unnecessary fright, so now, upon this notion spreading, viz., that the distemper was not so catching as formerly, and that if it was catched it was not so mortal, and seeing abundance of people who really fell sick recover again daily, they took to such a precipitant courage, and grew so entirely regardless of themselves and of the infection, that they made no more of the plague than of an ordinary fever, nor indeed so much. They not only went boldly into company with those who had tumours and carbuncles upon them that were running, and consequently contagious, but eat and drank with them, nay, into their houses to visit them, and even, as I was told, into their very chambers where they lay sick.

This I could not see rational. My friend Dr. Heath allowed, and it was plain to experience, that the disease was as catching as ever, and as many fell sick, but only he alledged that so many of those that fell sick did not die; but I think that while many did die, and that at best the distemper itself was very terrible, the sores and swellings very tormenting, and the danger of death not left out of the circumstances of sickness, though not so frequent as before; all those things, together with the exceeding tediousness of the cure, the loathsomeness of the disease, and many other articles, were enough to deter any man living from a dangerous mixture with the sick people, and make them as anxious almost to avoid the infection as before.

●   ●   ●

The consequence of this [an influx of visitors from out of town] was, that the bills increased again four hundred the very first week in November; and, if I might believe the physicians, there was above three thousand sick that week, most of them newcomers, too.

One John Cock . . . was an eminent example of this; I mean of the hasty return of the people when the plague was abated. This John Cock had left the town with his whole family, and locked up his house, and was gone in the country, as many others did; and finding the plague so decreased in November that there died but 905 per week of all diseases, he ventured home again. He had in his family ten persons; that is to say, himself and wife, five children, two apprentices, and a maidservant. He had not been returned to his house above a week, and began to open his shop and carry on his trade, but the distemper broke out in his family, and within about five days they all died, except one; that is to say, himself, his wife, all his five children, and his two apprentices; and only the maid remained alive.

Hawaii

James A. Michener

Actually, the question was of no importance, for Lahaina was about to be visited by a pestilence known as the scourge of the Pacific. On earlier trips to Hawaii this dreadful plague had wiped out more than half the population, and now it stood poised in the fo'c's'l of a whaler resting in Lahaina Roads, prepared to strike once more with demonic force, killing, laying waste, destroying an already doomed population. It was the worst disease of the Pacific: measles.

This time it started innocently by jumping from the diseased whaler and into the mission home, where immunities built up during a hundred generations in England and Massachusetts confined the disease to a trivial childhood sickness. Jerusha, inspecting her son Micah's chest one morning, found the customary red rash. "Have you a sore throat?" she asked, and when Micah said yes, she informed Abner, "I'm afraid our son has the measles."

Abner groaned and said, "I suppose Lucy and David and Esther are bound to catch it in turn," and he took down his medical books to see what he should do for the worrisome fever. Medication was simple and the routine not burdensome, so he said, "We'll plan for three weeks of keeping the children indoors." But it occurred to him that it might be prudent to see if John Whipple had any medicine for reducing the fever more quickly, and so he stopped casually by J & W's to report, "Worse luck! Micah seems to have the measles and I suppose . . ."

Whipple dropped his pen and cried, "Did you say measles?"

"Well, spots on his chest."

"Oh, my God!" Whipple mumbled, grabbing his bag and rushing to the mission house. With trembling fingers he inspected the sick boy and Jerusha saw that the doctor was perspiring.

"Are measles so dangerous?" she asked with apprehension.

"Not for him," Whipple replied. He then led the parents into the front room and asked in a whisper, "Have you been in contact with any Hawaiians since Micah became ill?"

"No," Abner reflected. "I walked down to your store."

"Thank God," Whipple gasped, washing his hands carefully. "Abner, we have only a slight chance of keeping this dreadful disease away from the Hawaiians, but I want your entire family to stay in this house for three weeks. See nobody."

Jerusha challenged him directly: "Brother John, is it indeed the measles?"

"It is," he replied, "and I would to God it were anything else. We had better prepare ourselves, for there may be sad days ahead." Then, awed by the gravity of the threat, he asked impulsively, "Abner, would you please say a prayer for all of us . . . for Lahaina? Keep the pestilence from this town." And they knelt while Abner prayed.

But men from the infected whaler had moved freely through the community, and on the next morning Dr. Whipple happened to look out of his door to see a native man, naked, digging himself a shallow grave beside the ocean, where cool water could seep in and fill the sandy rectangle. Rushing to the reef, Whipple called, "Kekuana, what are you doing?" And the Hawaiian, shivering fearfully, replied, "I am burning to death and the water will cool me." At this Dr. Whipple said sternly, "Go back to your home, Kekuana, and wrap yourself in tapa. Sweat this illness out or you will surely die." But the man argued, "You do not know how terrible the burning fire is," and he sank himself in the salt water and within the day he died.

Now all along the beach Hawaiians, spotted with measles, dug themselves holes in the cool wet sand, and in spite of anything Dr. Whipple could tell them, crawled into the comforting waters and died. The cool irrigation ditches and taro patches were filled with corpses. Through the miserable huts of the town the pestilence swept like fire, burning its victims with racking fevers that could not be endured. Dr. Whipple organized his wife, the Hales and the Janderses into a medical team that worked for three weeks, arguing, consoling and burying. Once Abner cried in frustration, "John, why do these stubborn people insist upon plunging into the surf when they know it kills them?" And Whipple replied in exhaustion, "We are misled because we call the fever measles. In these unprotected people it is something much worse. Abner, you have never known such a fever."

Nevertheless the little missionary pleaded with his patients, "If you go into the water, you will die."

"I want to die, Makua Hale," they replied.

Jerusha and Amanda saved many lives by forcing their way into huts where they took away babies without even asking, for they knew that if the fevered infants continued their piteous moaning their parents would carry them to the sea. By wrapping the children in blankets and dosing them with syrup of squill, thus encouraging the fever to erupt through skin sores, as it should, the women rescued the children, but with adults neither logic nor force could keep them from the sea, and throughout Lahaina one Hawaiian in three perished.

In time the measles reached even Malama's walled-in compound where it struck Keoki, who welcomed it, and his baby son Kelolo. Here the Hales found the shivering Kanakoa family, and Jerusha said promptly, "I will take the little boy home with me." And there must have been a great devil near Abner's heart, for when his wife had the dying child in her arms he stopped her and asked, "Would it not be better if that child of sin . . . ?"

Jerusha looked steadily at her husband and said, "I will take the boy. This is what we have been preaching about in the new laws — All the children." And she carried the whimpering child and placed him among her own.

When she was gone, Abner found that Keoki had escaped to the seashore where he dug a shallow grave into which salt water seeped, and before Abner could overtake him he had plunged in, finding relief at last. Abner, limping along the reef, came upon him and cried, "Keoki, if you do that you will surely die."

"I shall die," the tall alii shivered.

Compassionately, Abner pleaded, "Come back, and I will wrap you in blankets."

"I shall die," Keoki insisted.

"There is no evil that God cannot forgive," Abner assured the quaking man.

"Your God no longer exists," Keoki mumbled from his cold grave. "I shall die and renew my life in the waters of Kane."

Abner was horrified by these words, and pleaded, "Keoki, even in death do not use such blasphemy against the God who loves you."

"Your god brings us only pestilence," the shivering man replied.

"I am going to pray for you, Keoki."

"It's too late now. You never wanted me in your church," and the fever-racked alii splashed his face with water.

"Keoki!" Abner pleaded. "You are dying. Pray with me for your immortal soul."

"Kane will protect me," the stricken young man insisted.

"Oh, no! No!" Abner cried, but he felt a strong hand take his arm and pull him from the grave.

It was one-eyed Kelolo, who said, "You must leave my son alone with his god."

"No!" Abner shouted passionately. "Keoki, will you pray with me?"

"I am beginning a dark journey," the sick man replied feebly. "I have told Kane of my coming. No other prayers are necessary."

The incoming tide brought fresh and colder waters into the grave, and at that moment Abner leaped into the shallow pit and grasped his old friend by the hands. "Keoki, do not die in darkness. My dearest brother . . ." But the alii drew away from Abner and hid his parched face with his forearms.

"Take him away," the young man cried hoarsely, "I will die with my own god." And Kelolo dragged Abner from the grave.

When the pestilence was ended, Abner and Jerusha brought the baby Kelolo, now healthy and smiling, back to the palace, where Noelani took the child and studied it dispassionately. "This one will be the last of the alii," she predicted sadly. "But it may be better that way. Another pestilence and we will all be gone."

Minha Vida de Menina*

Elizabeth Bishop

Thursday, March 9, 1893

My father thought it funny when I said I was jealous of Luizinha who was out on the street with a handkerchief on her face.

I do not want a toothache because I have seen everyone cry so, that I think it must be terribly painful. Naninha, when she had a toothache, drove everyone crazy. Aunt Agostinha just prayed and made vows for fear Naninha would lose her mind. She screamed, rolled on the floor, and banged her head on the wall until one would think she were a mental patient. The other day she screamed so loudly that people came in from the street to help but she cursed them and rolled in the garden.

No one knows how to stop a toothache. Pull it out? No one thinks of this after what happened to Dona Augusta's daughter. She screamed many days with a toothache. Her father, discouraged, called the dentist to pull it out. He pulled it and the poor little thing had relief for only a little time for after that she died a horrible death: she stiffened all over, her teeth locked, and her head bent back until she died.

Luizinha had a screaming toothache this week. Mama had her rinse her mouth with salt water, put snuff on her tooth, put creosote on it, but nothing worked. It was Senhora Ritinha who cured her in a strange way. She gave her castor oil and on the next day her face swelled and she stopped crying. Today she asked to go out without the handkerchief on her face and Mama was horrified with the idea that her face would be ruined. I thought that it was because of this that Senhor Cula's Belinha went all her life with her face wrapped in a handkerchief. Today I wanted to go out with a handkerchief like I had seen the others doing but Mama wouldn't let me.

The Death of Ivan Ilych*

Leo Tolstoy

They were all in good health. It could not be called ill health if Ivan Ilych sometimes said that he had a queer taste in his mouth and felt some discomfort in his left side.

But this discomfort increased and, though not exactly painful, grew into a sense of pressure in his side accompanied by ill humour. And his irritability became worse and worse and began to mar the agreeable, easy, and correct life that had established itself in the Golovin family. Quarrels between husband and wife became more and more frequent, and soon the ease and amenity disappeared and even the decorum was barely maintained. Scenes again became frequent, and very few of those islets remained on which husband and wife could meet without an explosion. Praskovya Fëdorovna now had good reason to say that her husband's temper was trying. With characteristic exaggeration she said he had always had a dreadful temper, and that it had needed all her good nature to put up with it for twenty years. It was true that now the quarrels were started by him. His bursts of temper always came just before dinner, often just as he began to eat his soup. Sometimes he noticed that a plate or dish was chipped, or the food was not right, or his son put his elbow on the table, or his daughter's hair was not done as he liked it, and for all this he blamed Praskovya Fëdorovna. At first she retorted and said disagreeable things to him, but once or twice he fell into such a rage at the beginning of dinner that she realized it was due to some physical derangement brought on by taking food, and so she restrained herself and did not answer, but only hurried to get the dinner over. She regarded this self-restraint as highly praiseworthy. Having come to the conclusion that her husband had a dreadful temper and made her life miserable, she began to feel sorry for herself, and the more she pitied herself the more she hated her husband. She began to wish he would die; yet she did not want him to die because then his salary would cease. And this irritated her against him still more. She considered herself dreadfully unhappy just because not even his death could save her, and though she concealed her exasperation, that hidden exasperation of hers increased his irritation also.

*From The Death of Ivan Ilych by Leo Tolstoy, translated by Aylmer Maude and published by Oxford University Press.

After one scene in which Ivan Ilych had been particularly unfair and after which he had said in explanation that he certainly was irritable but that it was due to his not being well, she said that if he was ill it should be attended to, and insisted on his going to see a celebrated doctor.

He went. Everything took place as he had expected and as it always does. There was the usual waiting and the important air assumed by the doctor, with which he was so familiar (resembling that which he himself assumed in court), and the sounding and listening, and the questions which called for answers that were foregone conclusions and were evidently unnecessary, and the look of importance which implied that "if only you put yourself in our hands we will arrange everything — we know indubitably how it has to be done, always in the same way for everybody alike." It was all just as it was in the law courts. The doctor put on just the same air towards him as he himself put on towards an accused person.

The doctor said that so-and-so indicated that there was so-and-so inside the patient, but if the investigation of so-and-so did not confirm this, then he must assume that and that. If he assumed that and that, then . . . and so on. To Ivan Ilych only one question was important: was his case serious or not? But the doctor ignored that inappropriate question. From his point of view it was not the one under consideration, the real question was to decide between a floating kidney, chronic catarrh, or appendicitis. It was not a question of Ivan Ilych's life or death, but one between a floating kidney and appendicitis. And that question the doctor solved brilliantly, as it seemed to Ivan Ilych, in favour of the appendix, with the reservation that should an examination of the urine give fresh indications the matter would be reconsidered. All this was just what Ivan Ilych had himself brilliantly accomplished a thousand times in dealing with men on trial. The doctor summed up just as brilliantly, looking over his spectacles triumphantly and even gaily at the accused. From the doctor's summing up Ivan Ilych concluded that things were bad, but that for the doctor, and perhaps for everybody else, it was a matter of indifference, though for him it was bad. And this conclusion struck him painfully, arousing in him a great feeling of pity for himself and of bitterness towards the doctor's indifference to a matter of such importance.

He said nothing of this, but rose, placed the doctor's fee on the table, and remarked with a sigh: "We sick people probably often put inappropriate questions. But tell me, in general, is this complaint dangerous, or not? . . ."

The doctor looked at him sternly over his spectacles with one eye, as if to say: "Prisoner, if you will not keep to the questions put to you, I shall be obliged to have you removed from the court."

"I have already told you what I consider necessary and proper. The analysis may show something more." And the doctor bowed.

Ivan Ilych went out slowly, seated himself disconsolately in his sledge, and drove home. All the way home he was going over what the doctor had said, trying to translate those complicated, obscure, scientific phrases into plain language and find in them an answer to the question: "Is my condition bad? Is it very bad? Or is there as yet nothing much wrong?" And it seemed to him that the meaning of what the doctor had said was that it was very bad. Everything in the streets seemed depressing. The cabmen, the houses, the passers-by, and the shops, were dismal. His ache, this dull gnawing ache that never ceased for a moment, seemed to have acquired a new and more serious significance from the doctor's dubious remarks. Ivan Ilych now watched it with a new and oppressive feeling.

He reached home and began to tell his wife about it. She listened, but in the middle of his account his daughter came in with her hat on, ready to go out with her mother. She sat down reluctantly to listen to this tedious story, but could not stand it long, and her mother too did not hear him to the end.

"Well, I am very glad," she said. "Mind now to take your medicine regularly. Give me the prescription and I'll send Gerasim to the chemist's." And she went to get ready to go out.

While she was in the room Ivan Ilych had hardly taken time to breathe, but he sighed deeply when she left it.

"Well," he thought, "perhaps it isn't so bad after all."

He began taking his medicine and following the doctor's directions, which had been altered after the examination of the urine. But then it happened that there was a contradiction, between the indications drawn from the examination of the urine and the symptoms that showed themselves. It turned out that what was happening differed from what the doctor had told him, and that he had either forgotten, or blundered, or hidden something from him. He could not, however, be blamed for that, and Ivan Ilych still obeyed his orders implicitly and at first derived some comfort from doing so.

From the time of his visit to the doctor, Ivan Ilych's chief occupation was the exact fulfilment of the doctor's instructions regarding hygiene and the taking of medicine, and the observation of his pain and his excretions. His chief interests came to be people's ailments and people's health. When sickness, deaths, or recoveries were mentioned in his presence, especially when the illness resembled his own, he listened with agitation which he tried to hide, asked questions, and applied what he heard to his own case.

The pain did not grow less, but Ivan Ilych made efforts to force himself to think that he was better. And he could do this so long as nothing agitated him. But as soon as he had any unpleasantness with his wife, any lack of success in his official work, or held bad cards at bridge, he was at once acutely sensible of his disease. He had formerly borne such mischances, hoping soon to adjust what was wrong, to master it and attain success, or make a grand slam. But now every mischance upset him and plunged him into despair. He would say to himself: "There now, just as I was beginning to get better and the medicine had begun to take effect, comes this accursed misfortune, or unpleasantness. . . ." And he was furious with the mishap, or with the people who were causing the unpleasantness and killing him, for he felt that this fury was killing him but could not restrain it. One would have thought that it should have been clear to him that this exasperation with circumstances and people aggravated his illness, and that he ought therefore to ignore unpleasant occurrences. But he drew the very opposite conclusion: he said that he needed peace, and he watched for everything that might disturb it and became irritable at the slightest infringement of it. His condition was rendered worse by the fact that he read medical books and consulted doctors. The progress of his disease was so gradual that he could deceive himself when comparing one day with another — the difference was so slight. But when he consulted the doctors it seemed to him that he was getting worse, and even very rapidly. Yet despite this he was continually consulting them.

That month he went to see another celebrity, who told him almost the same as the first had done but put his questions rather differently, and the interview with this celebrity only increased Ivan Ilych's doubts and fears. A friend of a friend of his, a very good doctor, diagnosed his illness again quite differently from the others, and though he predicted recovery, his questions and suppositions bewildered Ivan Ilych still more and increased his doubts. A homeopathist diagnosed the disease in yet another way, and prescribed medicine which Ivan Ilych took secretly for a week. But after a week, not feeling any improvement and having lost confidence both in the former doctor's treatment and in this one's, he became still more despondent. One day a lady acquaintance mentioned a cure effected by a wonder-working icon. Ivan Ilych caught himself listening attentively and beginning to believe that it had occurred. This incident alarmed him. "Has my mind really weakened to such an extent?" he asked himself. "Nonsense! It's all rubbish. I mustn't give way to nervous fears but having chosen a doctor must keep strictly to his treatment. That is what I will do. Now it's all settled. I won't think about it, but will follow the treatment seriously till summer, and then we shall see. From now there must be no more of this wavering!" This was easy to say but impossible to carry out. The pain in his side oppressed him and seemed to grow worse and more incessant, while the taste in his mouth grew stranger and stranger. It seemed to him that his breath had a disgusting smell, and he was conscious of a loss of appetite and strength. There was no deceiving himself: something terrible, new, and more important than anything before in his life, was taking place within him of which he alone was aware. Those about him did not understand or would not understand it, but thought everything in the world was going on as usual. That tormented Ivan Ilych more than anything. He saw that his household, especially his wife and daughter who were in a perfect whirl of visiting, did not understand anything of it and were annoyed that he was so depressed and so exacting, as if he were to blame for it. Though they tried to disguise it he saw that he was an obstacle in their path, and that his wife had adopted a definite line in regard to his illness and kept to it regardless of anything he said or did. Her attitude was this: "You know," she would say to her friends, "Ivan Ilych can't do as other people do, and keep to the treatment prescribed for him. One day he'll take his drops and keep strictly to his diet and go to bed in good time, but the next day unless I watch him he'll suddenly forget his medicine, eat sturgeon — which is forbidden — and sit up playing cards till one o'clock in the morning."

"Oh, come, when was that?" Ivan Ilych would ask in vexation. "Only once at Peter Ivanovich's."

"And yesterday with Shebek."

"Well, even if I hadn't stayed up, this pain would have kept me awake."

"Be that as it may you'll never get well like that, but will always make us wretched."

Praskovya Fëdorovna's attitude to Ivan Ilych's illness as she expressed it both to others and to him, was that it was his own fault and was another of the annoyances he caused her. Ivan Ilych felt that his opinion escaped her involuntarily — but that did not make it easier for him.

At the law courts too, Ivan Ilych noticed, or thought he noticed, a strange attitude towards himself. It sometimes seemed to him that people were watching him inquisitively as a man whose place might soon be vacant. Then again, his friends would suddenly begin to chaff him in a friendly way about his low spirits, as if the awful, horrible, and unheard-of thing that was going on within him, incessantly gnawing at him and irresistibly drawing him away, was a very agreeable subject for jests. Schwartz in particular irritated him by his jocularity, vivacity, and savoir-faire, which reminded him of what he himself had been ten years ago.

Friends came to make up a set and they sat down to cards. They dealt, bending the new cards to soften them, and he sorted the diamonds in his hand and found he had seven. His partner said "No trumps" and supported him with two diamonds. What more could be wished for? It ought to be jolly and lively. They would make a grand slam. But suddenly Ivan Ilych was conscious of that gnawing pain, that taste in his mouth, and it seemed ridiculous that in such circumstances he should be pleased to make a grand slam.

He looked at his partner Mikhail Mikhaylovich, who rapped the table with his strong hand and instead of snatching up the tricks pushed the cards courteously and indulgently towards Ivan Ilych that he might have the pleasure of gathering them up without the trouble of stretching out his hand for them. "Does he think I am too weak to stretch out my arm?" thought Ivan Ilych, and forgetting what he was doing he over-trumped his partner, missing the grand slam by three tricks. And what was most awful of all was that he saw how upset Mikhail Mikhaylovich was about it but did not himself care. And it was dreadful to realize why he did not care.

They all saw that he was suffering, and said: "We can stop if you are tired. Take a rest." Lie down? No, he was not at all tired, and he finished the rubber.

All were gloomy and silent. Ivan Ilych felt that he had diffused this gloom over them and could not dispel it. They had supper and went away, and Ivan Ilych was left alone with the consciousness that his life was poisoned and was poisoning the lives of others, and that this poison did not weaken but penetrated more and more deeply into his whole being.

With this consciousness, and with physical pain besides the terror, he must go to bed, often to lie awake the greater part of the night. Next morning he had to get up again, dress, go to the law courts, speak, and write; or if he did not go out, spend at home those twenty-four hours a day each of which was a torture. And he had to live thus all alone on the brink of an abyss, with no one who understood or pitied him.

So one month passed and then another. Just before the New Year his brother-in-law came to town and stayed at their house. Ivan Ilych was at the law courts and Praskovya Fëdorovna had gone shopping. When Ivan Ilych came home and entered his study he found his brother-in-law there — a healthy, florid man — unpacking his portmanteau himself. He raised his head on hearing Ivan Ilych's footsteps and looked up at him for a moment without a word. That stare told Ivan Ilych everything. His brother-in-law opened his mouth to utter an exclamation of surprise but checked himself, and that action confirmed it all.

"I have changed, eh?"

"Yes, there is a change."

And after that, try as he would to get his brother-in-law to return to the subject of his looks, the latter would say nothing about it. Praskovya Fëdorovna came home and her brother went out to her. Ivan Ilych locked the door and began to examine himself in the glass, first full face, then in profile. He took up a portrait of himself taken with his wife, and compared it with that he saw in the glass. The change in him was immense. Then he bared his arms to the elbow, looked at them, drew the sleeves down again, sat down on an ottoman, and grew blacker than night.

"No, no, this won't do!" he said to himself, and jumped up, went to the table, took up some law papers and began to read them, but could not continue. He unlocked the door and went into the reception-room. The door leading to the drawing-room was shut. He approached it on tiptoe and listened.

"No, you are exaggerating!" Praskovya Fëdorovna was saying.

"Exaggerating! Don't you see it? Why, he's a dead man! Look at his eyes — there's no light in them. But what is it that is wrong with him?"

"No one knows. Nikolaevich [that was another doctor] said something, but I don't know what. And Leshchetitsky [this was the celebrated specialist] said quite the contrary . . ."

Ivan Ilych walked away, went to his room, lay down, and began musing: "The kidney, a floating kidney." He recalled all the doctors had told him of how it detached itself and swayed about. And by an effort of imagination he tried to catch that kidney and arrest it and support it. So little was needed for this, it seemed to him. "No, I'll go to see Peter Ivanovich again." [That was the friend whose friend was a doctor.] He rang, ordered the carriage, and got ready to go.

"Where are you going, Jean?" asked his wife, with a specially sad and exceptionally kind look.

This exceptionally kind look irritated him. He looked morosely at her.

"I must go to see Peter Ivanovich."

He went to see Peter Ivanovich, and together they went to see his friend, the doctor. He was in, and Ivan Ilych had a long talk with him.

Reviewing the anatomical and physiological details of what in the doctor's opinion was going on inside him, he understood it all.

There was something, a small thing, in the vermiform appendix. It might all come right. Only stimulate the energy of one organ and check the activity of another, then absorption would take place and everything would come right. He got home rather late for dinner, ate his dinner, and conversed cheerfully, but could not for a long time bring himself to go back to work in his room. At last, however, he went to his study and did what was necessary, but the consciousness that he had put something aside — an important, intimate matter which he would revert to when his work was done — never left him. When he had finished his work he remembered that his intimate matter was the thought of his vermiform appendix. But he did not give himself up to it, and went to the drawing-room for tea. There were callers there, including the examining magistrate who was a desirable match for his daughter, and they were conversing, playing the piano, and singing. Ivan Ilych, as Praskovya Fëdorovna remarked, spent that evening more cheerfully than usual, but he never for a moment forgot that he had postponed the important matter of the appendix. At eleven o'clock he said good-night and went to his bedroom. Since his illness he had slept alone in a small room next to his study. He undressed and took up a novel by Zola, but instead of reading it he fell into thought, and in his imagination that desired improvement in the vermiform appendix occurred. There was the absorption and evacuation and the re-establishment of normal activity. "Yes, that's it!" he said to himself. "One need only assist nature, that's all." He remembered his medicine, rose, took it, and lay down on his back watching for the beneficent action of the medicine and for it to lessen the pain. "I need only take it regularly and avoid all injurious influences. I am already feeling better, much better." He began touching his side: it was not painful to the touch. "There, I really don't feel it. It's much better already." He put out the light and turned on his side. . . . "The appendix is getting better, absorption is occurring." Suddenly he felt the old, familiar, dull, gnawing pain, stubborn and serious. There was the same familiar loathsome taste in his mouth. His heart sank and he felt dazed. "My God! My God!" he muttered. "Again, again! and it will never cease." And suddenly the matter presented itself in a quite different aspect. "Vermiform appendix! Kidney!" he said to himself. "It's not a question of appendix or kidney, but of life and . . . death. Yes, life was there and now it is going, going and I cannot stop it. Yes. Why deceive myself? Isn't it obvious to everyone but me that I'm dying, and that it's only a question of weeks, days . . . it may happen this moment. There was light and now there is darkness. I was here and now I'm going there! Where?" A chill came over him, his breathing ceased, and he felt only the throbbing of his heart.

"When I am not, what will there be? There will be nothing. Then where shall I be when I am no more? Can this be dying? No, I don't want to!" He jumped up and tried to light the candle, felt for it with trembling hands, dropped candle and candlestick on the floor, and fell back on his pillow.

"What's the use? It makes no difference," he said to himself, staring with wide-open eyes into the darkness. "Death. Yes, death. And none of them know or wish to know it, and they have no pity for me. Now they are playing." (He heard through the door the distant sound of a song and its accompaniment.) "It's all the same to them, but they will die too! Fools! I first, and they later, but it will be the same for them. And now they are merry . . . the beasts!"

Anger choked him and he was agonizingly, unbearably miserable. "It is impossible that all men have been doomed to suffer this awful horror!" He raised himself.

"Something must be wrong. I must calm myself — must think it all over from the beginning." And he again began thinking. "Yes, the beginning of my illness: I knocked my side, but I was still quite well that day and the next. It hurt a little, then rather more.

●   ●   ●

So he departed, and the cheerful state of mind induced by his success and by the harmony between his wife and himself, the one intensifying the other, did not leave him. He found a delightful house, just the thing both he and his wife had dreamt of. Spacious, lofty reception rooms in the old style, a convenient and dignified study, rooms for his wife and daughter, a study for his son — it might have been specially built for them. Ivan Ilych himself super-intended the arrangements, chose the wallpapers, supplemented the furniture (preferably with antiques which he considered particularly comme il faut), and supervised the upholstering. Everything progressed and progressed and approached the ideal he had set himself; even when things were only half completed they exceeded his expectations. He saw what a refined and elegant character, free from vulgarity, it would all have when it was ready. On falling asleep he pictured to himself how the reception-room would look. Looking at the yet unfinished drawing-room he could see the fireplace, the screen, the what-not, the little chairs dotted here and there, the dishes and plates on the walls, and the bronzes, as they would be when everything was in place. He was pleased by the thought of how his wife and daughter, who shared his taste in this matter, would be impressed by it. They were certainly not expecting as much.

He had been particularly successful in finding, and buying cheaply, antiques which gave a particularly aristocratic character to the whole place. But in his letters he intentionally understated everything in order to be able to surprise them. All this so absorbed him that his new duties — though he liked his official work — interested him less than he had expected. Sometimes he even had moments of absent-mindedness during the Court Sessions, and would consider whether he should have straight or curved cornices for his curtains. He was so interested in it all that he often did things himself, rearranging the furniture, or rehanging the curtains. Once when mounting a step-ladder to show the upholsterer, who did not understand, how he wanted the hangings draped, he made a false step and slipped, but being a strong and agile man he clung on and only knocked his side against the knob of the window frame. The bruised place was painful but the pain soon passed, and he felt particularly bright and well just then. He wrote: "I feel fifteen years younger." He thought he would have everything ready by September, but it dragged on till mid-October. But the result was charming not only in his eyes but to everyone who saw it.

•   •   •

I saw the doctors, then followed despondency and anguish, more doctors, and I drew nearer to the abyss. My strength grew less and I kept coming nearer and nearer, and now I have wasted away and there is no light in my eyes. I think of the appendix — but this is death! I think of mending the appendix, and all the while here is death! Can it really be death?" Again terror seized him and he gasped for breath. He leant down and began feeling for the matches, pressing with his elbow on the stand beside the bed. It was in his way and hurt him, he grew furious with it, pressed on it still harder, and upset it. Breathless and in despair he fell on his back, expecting death to come immediately.

Meanwhile the visitors were leaving. Praskovya Fëdorovna was seeing them off. She heard something fall and came in.

"What has happened?"

"Nothing. I knocked it over accidentally."

She went out and returned with a candle. He lay there panting heavily, like a man who has run a thousand yards, and stared upwards at her with a fixed look.

"What is it, Jean?"

"No . . . o . . . thing. I upset it." ("Why speak of it? She won't understand," he thought.)

And in truth she did not understand. She picked up the stand, lit his candle, and hurried away to see another visitor off. When she came back he still lay on his back, looking upwards.

"What is it? Do you feel worse?"

"Yes."

She shook her head and sat down.

"Do you know, Jean, I think we must ask Leshchetitsky to come and see you here."

This meant calling in the famous specialist, regardless of expense. He smiled malignantly and said "No." She remained a little longer and then went up to him and kissed his forehead.

While she was kissing him he hated her from the bottom of his soul and with difficulty refrained from pushing her away.

"Good-night. Please God you'll sleep."

"Yes."

Ivan Ilych saw that he was dying, and he was in continual despair.

In the depth of his heart he knew he was dying, but not only was he not accustomed to the thought, he simply did not and could not grasp it.

The syllogism he had learnt from Kiezewetter's Logic: "Caius is a man, men are mortal, therefore Caius is mortal," had always seemed to him correct as applied to Caius, but certainly not as applied to himself. That Caius — man in the abstract — was mortal, was perfectly correct, but he was not Caius, not an abstract man, but a creature quite, quite separate from all others. He had been little Vanya, with a mamma and a papa, with Mitya and Volodya, with the toys, a coachman and a nurse, afterwards with Katenka and with all the joys, griefs, and delights of childhood, boyhood, and youth. What did Caius know of the smell of that striped leather ball Vanya had been so fond of? Had Caius kissed his mother's hand like that, and did the silk of her dress rustle so for Caius? Had he rioted like that at school when the pastry was bad? Had Caius been in love like that? Could Caius preside at a session as he did? "Caius really was mortal, and it was right for him to die? but for me, little Vanya, Ivan Ilych, with all my thoughts and emotions, it's altogether a different matter. It cannot be that I ought to die. That would be too terrible."

Such was his feeling.

"If I had to die like Caius I should have known it was so. An inner voice would have told me so, but there was nothing of the sort in me and I and all my friends felt that our case was quite different from that of Caius. And now here it is!" he said to himself. "It can't be. It's impossible! But here it is. How is this? How is one to understand it?"

He could not understand it, and tried to drive this false, incorrect, morbid thought away and to replace it by other proper and healthy thoughts. But that thought, and not the thought only but the reality itself, seemed to come and confront him.

And to replace that thought he called up a succession of others, hoping to find in them some support. He tried to get back into the former current of thoughts that had once screened the thought of death from him. But strange to say, all that had formerly shut off, hidden, and destroyed, his consciousness of death, no longer had that effect. Ivan Ilych now spent most of his time in attempting to re-establish that old current. He would say to himself: "I will take up my duties again — after all I used to live by them." And banishing all doubts he would go to the law courts, enter into conversation with his colleagues, and sit carelessly as was his wont, scanning the crowd with a thoughtful look and leaning both his emaciated arms on the arms of his oak chair; bending over as usual to a colleague and drawing his papers nearer he would interchange whispers with him, and then suddenly raising his eyes and sitting erect would pronounce certain words and open the proceedings. But suddenly in the midst of those proceedings the pain in his side, regardless of the stage the proceedings had reached, would begin its own gnawing work. Ivan Ilych would turn his attention to it and try to drive the thought of it away, but without success. IT would come and stand before him and look at him, and he would be petrified and the light would die out of his eyes, and he would again begin asking himself whether IT alone was true. And his colleagues and subordinates would see with surprise and distress that he, the brilliant and subtle judge, was becoming confused and making mistakes. He would shake himself, try to pull himself together, manage somehow to bring the sitting to a close, and return home with the sorrowful consciousness that his judicial labours could not as formerly hide from him what he wanted them to hide, and could not deliver him from IT. And what was worst of all was that IT drew his attention to itself not in order to make him take some action but only that he should look at IT, look it straight in the face: look at it and without doing anyting, suffer inexpressibly.

And to save himself from this condition Ivan Ilych looked for consolations — new screens — and new screens were found and for a while seemed to save him, but then they immediately fell to pieces or rather became transparent, as if IT penetrated them and nothing could veil IT.

In these latter days he would go into the drawing-room he had arranged — that drawing-room where he had fallen and for the sake of which (how bitterly ridiculous it seemed) he had sacrificed his life — for he knew that his illness originated with that knock. He would enter and see that something had scratched the polished table. He would look for the cause of this and find that it was the bronze ornamentation of an album, that had got bent. He would take up the expensive album which he had lovingly arranged, and feel vexed with his daughter and her friends for their untidiness — for the album was torn here and there and some of the photographs turned upside down. He would put it carefully in order and bend the ornamentation back into position. Then it would occur to him to place all those things in another corner of the room, near the plants. He could call the footman, but his daughter or wife would come to help him. They would not agree, and his wife would contradict him, and he would dispute and grow angry. But that was all right, for then he did not think about IT. IT was invisible.

But then, when he was moving something himself, his wife would say: "Let the servants do it. You will hurt yourself again." And suddenly IT would flash through the screen and he would see it. It was just a flash, and he hoped it would disappear, but he would involuntarily pay attention to his side. "It sits there as before, gnawing just the same!" And he could no longer forget IT, but could distinctly see it looking at him from behind the flowers. "What is it all for?"

"It really is so! I lost my life over that curtain as I might have done when storming a fort. Is that possible? How terrible and how stupid. It can't be true! It can't, but it is."

He would go to his study, lie down, and again be alone with IT face to face with IT. And nothing could be done with IT except to look at it and shudder.

How it happened it is impossible to say because it came about step by step, unnoticed, but in the third month of Ivan Ilych's illness, his wife, his daughter, his son, his acquaintances, the doctors, the servants, and above all he himself, were aware that the whole interest he had for other people was whether he would soon vacate his place, and at last release the living from the discomfort caused by his presence and be himself released from his sufferings.

He slept less and less. He was given opium and hypodermic injections of morphine, but this did not relieve him. The dull depression he experienced in a somnolent condition at first gave him a little relief, but only as something new, afterwards it became as distressing as the pain itself or even more so.

Special foods were prepared for him by the doctors' orders, but all those foods became increasingly distasteful and disgusting to him.

For his excretions also special arrangements had to be made, and this was a torment to him every time — a torment from the uncleanliness, the unseemliness, and the smell, and from knowing that another person had to take part in it.

But just through this most unpleasant matter, Ivan Ilych obtained comfort. Gerasim, the butler's young assistant, always came in to carry the things out. Gerasim was a clean, fresh peasant lad, grown stout on town food and always cheerful and bright. At first the sight of him, in his clean Russian peasant costume, engaged in that disgusting task embarrassed Ivan Ilych.

Once when he got up from the commode too weak to draw up his trousers, he dropped into a soft armchair and looked with horror at his bare, enfeebled thighs with the muscles so sharply marked on them.

Gerasim with a firm light tread, his heavy boots emitting a pleasant smell of tar and fresh winter air, came in wearing a clean Hessian apron, the sleeves of his print shirt tucked up over his strong bare young arms; and refraining from looking at his sick master out of consideration for his feelings, and restraining the joy of life that beamed from his face, he went up to the commode.

"Gerasim!" said Ivan Ilych in a weak voice.

Gerasim started, evidently afraid he might have committed some blunder, and with a rapid movement turned his fresh, kind, simple young face which just showed the first downy signs of a beard.

"Yes, sir?"

"That must be very unpleasant for you. You must forgive me. I am helpless."

"Oh, why, sir," and Gerasim's eyes beamed and he showed his glistening white teeth, "what's a little trouble? It's a case of illness with you, sir."

And his deft strong hands did their accustomed task, and he went out of the room stepping lightly. Five minutes later he as lightly returned.

Ivan Ilych was still sitting in the same position in the armchair.

"Gerasim," he said then the latter had replaced the freshly-washed utensil. "Please come here and help me." Gerasim went up to him. "Lift me up. It is hard for me to get up, and I have sent Dmitri away."

Gerasim went up to him, grasped his master with his strong arms deftly but gently, in the same way that he stepped — lifted him, supported him with one hand, and with the other drew up his trousers and would have set him down again, but Ivan Ilych asked to be led to the sofa. Gerasim, without an effort and without apparent pressure, led him, almost lifting him, to the sofa and placed him on it.

"Thank you. How easily and well you do it all!"

Gerasim smiled again and turned to leave the room. But Ivan Ilych felt his presence such a comfort that he did not want to let him go.

"One thing more, please move up that chair. No, the other one — under my feet. It is easier for me when my feet are raised."

Gerasim brought the chair, set it down gently in place, and raised Ivan Ilych's legs on to it. It seemed to Ivan Ilych that he felt better while Gerasim was holding up his legs.

"It's better when my legs are higher," he said. "Place that cushion under them."

Gerasim did so. He again lifted the legs and placed them, and again Ivan Ilych felt better while Gerasim held his legs. When he set them down Ivan Ilych fancied he felt worse.

"Gerasim," he said. "Are you busy now?"

"Not at all, sir," said Gerasim, who had learnt from the townsfolk how to speak to gentlefolk.

"What have you still to do?"

"What have I to do? I've done everything except chopping the logs for tomorrow."

"Then hold my legs up a bit higher, can you?"

"Of course I can. Why not?" And Gerasim raised his master's legs higher and Ivan Ilych thought that in that position he did not feel any pain at all.

"And how about the logs?"

"Don't trouble about that, sir. There's plenty of time."

Ivan Ilych told Gerasim to sit down and hold his legs, and began to talk to him. And strange to say it seemed to him that he felt better while Gerasim held his legs up.

After that Ivan Ilych would sometimes call Gerasim and get him to hold his legs on his shoulders, and he liked talking to him. Gerasim did it all easily, willingly, simply, and with a good nature that touched Ivan Ilych. Health, strength, and vitality in other people were offensive to him, but Gerasim's strength and vitality did not mortify but soothed him.

What tormented Ivan Ilych most was the deception, the lie, which for some reason they all accepted, that he was not dying but was simply ill, and that he only need keep quiet and undergo a treatment and then something very good would result. He however knew that do what they would nothing would come of it, only still more agonizing suffering and death. This deception tortured him — their not wishing to admit what they all knew and what he knew, but wanting to lie to him concerning his terrible condition, and wishing and forcing him to participate in that lie. Those lies — lies enacted over him on the eve of his death and destined to degrade this awful, solemn act to the level of their visitings, their curtains, their sturgeon for dinner — were a terrible agony for Ivan Ilych. And strangely enough, many times when they were going through their antics over him he had been within a hairbreadth of calling out to them: "Stop lying! You know and I know that I am dying. Then at least stop lying about it!" But he had never had the spirit to do it. The awful, terrible act of his dying was, he could see, reduced by those about him to the level of a casual, unpleasant, and almost indecorous incident (as if someone entered a drawing-room diffusing an unpleasant odour) and this was done by that very decorum which he had served all his life long. He saw that no one felt for him, because no one even wished to grasp his position. Only Gerasim recognized it and pitied him. And so Ivan Ilych felt at ease only with him. He felt comforted when Gerasim supported his legs (sometimes all night long) and refused to go to bed, saying: "Don't you worry, Ivan Ilych. I'll get sleep enough later on," or when he suddenly became familiar and exclaimed: "If you weren't sick it would be another matter, but as it is, why should I grudge a little trouble?" Gerasim alone did not lie; everything showed that he alone understood the facts of the case and did not consider it necessary to disguise them, but simply felt sorry for his emaciated and enfeebled master. Once when Ivan Ilych was sending him away he even said straight out: "We shall all of us die, so why should I grudge a little trouble?" — expressing the fact that he did not think his work burdensome, because he was doing it for a dying man and hoped someone would do the same for him when his time came.

Apart from this lying, or because of it, what most tormented Ivan Ilych was that no one pitied him as he wished to be pitied. At certain moments after prolonged suffering he wished most of all (though he would have been ashamed to confess it) for someone to pity him as a sick child is pitied. He longed to be petted and comforted. He knew he was an important functionary, that he had a beard turning grey, and that therefore what he longed for was impossible, but still he longed for it. And in Gerasim's attitude towards him there was something akin to what he wished for, and so that attitude comforted him. Ivan Ilych wanted to weep, wanted to be petted and cried over, and then his colleague Shebek would come, and instead of weeping and being petted, Ivan Ilych would assume a serious, severe, and profound air, and by force of habit would express his opinion on a decision of the Court of Cassation and would stubbornly insist on that view. This falsity around him and within him did more than anything else to poison his last days.

It was morning. He knew it was morning because Gerasim had gone, and Peter the footman had come and put out the candles, drawn back one of the curtains, and begun quietly to tidy up. Whether it was morning or evening, Friday or Sunday, made no difference, it was all just the same: the gnawing, unmitigated, agonizing pain, never ceasing for an instant, the consciousness of life inexorably waning but not yet extinguished, the approach of that ever dreaded and hateful Death which was the only reality, and always the same falsity. What were days, weeks, hours, in such a case?

"Will you have some tea, sir?"

"He wants things to be regular, and wishes the gentlefolk to drink tea in the morning," thought Ivan Ilych, and only said "No."

"Wouldn't you like to move onto the sofa, sir?"

"He wants to tidy up the room, and I'm in the way. I am uncleanliness and disorder," he thought, and said only:

"No, leave me alone."

The man went on bustling about. Ivan Ilych stretched out his hand. Peter came up, ready to help.

"What is it, sir?"

"My watch."

Peter took the watch which was close at hand and gave it to his master.

"Half-past eight. Are they up?"

"No, sir, except Vladimir Ivanich" (the son) "who has gone to school. Praskovya Fëdorovna ordered me to wake her if you asked for her. Shall I do so?"

"No, there's no need to." "Perhaps I'd better have some tea," he thought, and added aloud: "Yes, bring me some tea."

Peter went to the door, but Ivan Ilych dreaded being left alone. "How can I keep him here? Oh yes, my medicine." "Peter, give me my medicine." "Why not? Perhaps it may still do me some good." He took a spoonful and swallowed it. "No, it won't help. It's all tomfoolery, all deception," he decided as soon as he became aware of the familiar, sickly, hopeless taste. "No, I can't believe in it any longer. But the pain, why this pain? If it would only cease just for a moment!" And he moaned. Peter turned towards him. "It's all right. Go and fetch me some tea."

Peter went out. Left alone Ivan Ilych groaned not so much with pain, terrible though that was, as from mental anguish. Always and for ever the same, always these endless days and nights. If only it would come quicker! If only what would come quicker? Death, darkness? . . . No, no! Anything rather than death!

When Peter returned with the tea on a tray, Ivan Ilych stared at him for a time in perplexity, not realizing who and what he was. Peter was disconcerted by that look and his embarrassment brought Ivan Ilych to himself.

"Oh, tea! All right, put it down. Only help me to wash and put on a clean shirt."

And Ivan Ilych began to wash. With pauses for rest, he washed his hands and then his face, cleaned his teeth, brushed his hair, and looked in the glass. He was terrified by what he saw, especially by the limp way in which his hair clung to his pallid forehead.

While his shirt was being changed he knew that he would be still more frightened at the sight of his body, so he avoided looking at it. Finally he was ready. He drew on a dressing-gown, wrapped himself in a plaid, and sat down in the armchair to take his tea. For a moment he felt refreshed, but as soon as he began to drink the tea he was again aware of the same taste, and the pain also returned. He finished it with an effort, and then lay down stretching out his legs, and dismissed Peter.

Always the same. Now a spark of hope flashes up, then a sea of despair rages, and always pain; always pain, always despair, and always the same. When alone he had a dreadful and distressing desire to call someone, but he knew beforehand that with others present it would be still worse. "Another dose of morphine — to lose consciousness. I will tell him, the doctor, that he must think of something else. It's impossible, impossible, to go on like this."

An hour and another pass like that. But now there is a ring at the door bell. Perhaps it's the doctor? It is. He comes in fresh, hearty, plump, and cheerful, with that look on his face that seems to say: "There now, you're in a panic about something, but we'll arrange it all for you directly!" The doctor knows this expression is out of place here, but he has put it on once for all and can't take it off — like a man who has put on a frock-coat in the morning to pay a round of calls.

The doctor rubs his hands vigorously and reassuringly.

"Brr! How cold it is! There's such a sharp frost; just let me warm myself!" he says, as if it were only a matter of waiting till he was warm, and then he would put everything right.

"Well now, how are you?"

Ivan Ilych feels that the doctor would like to say: "Well, how are our affairs?" but that even he feels that this would not do, and says instead: "What sort of a night have you had?"

Ivan Ilych looks at him as much as to say: "Are you really never ashamed of lying?" But the doctor does not wish to understand this question, and Ivan Ilych says: "Just as terrible as ever. The pain never leaves me and never subsides. If only something . . ."

"Yes, you sick people are always like that. . . . There, now I think I am warm enough. Even Praskovya Fëdorovna, who is so particular, could find no fault with my temperature. Well, now I can say good-morning," and the doctor presses his patient's hand.

Then, dropping his former playfulness, he begins with a most serious face to examine the patient, feeling his pulse and taking his temperature, and then begins the sounding and auscultation.

Ivan Ilych knows quite well and definitely that all this is nonsense and pure deception, but when the doctor, getting down on his knee, leans over him, putting his ear first higher then lower, and performs various gymnastic movements over him with a significant expression on his face, Ivan Ilych submits to it all as he used to submit to the speeches of the lawyers, though he knew very well that they were all lying and why they were lying.

The doctor, kneeling on the sofa, is still sounding him when Praskovya Fëdorovna's silk dress rustles at the door and she is heard scolding Peter for not having let her know of the doctor's arrival.

She comes in, kisses her husband, and at once proceeds to prove that she has been up a long time already, and only owing to a misunderstanding failed to be there when the doctor arrived.

Ivan Ilych looks at her, scans her all over, sets against her the whiteness and plumpness and cleanness of her hands and neck, the gloss of her hair, and the sparkle of her vivacious eyes. He hates her with his whole soul. And the thrill of hatred he feels for her makes him suffer from her touch.

Her attitude towards him and his disease is still the same. Just as the doctor had adopted a certain relation to his patient which he could not abandon, so had she formed one towards him — that he was not doing something he ought to do and was himself to blame, and that she reproached him lovingly for this — and she could not now change that attitude.

"You see he doesn't listen to me and doesn't take his medicine at the proper time. And above all he lies in a position that is no doubt bad for him — with his legs up."

She described how he made Gerasim hold his legs up.

The doctor smiled with a contemptuous affability that said: "What's to be done? These sick people do have foolish fancies of that kind, but we must forgive them."

When the examination was over the doctor looked at his watch, and then Praskovya Fëdorovna announced to Ivan Ilych that it was of course as he pleased, but she had sent today for a celebrated specialist who would examine him and have a consultation with Michael Danilovich (their regular doctor).

"Please don't raise any objections. I am doing this for my own sake," she said ironically, letting it be felt that she was doing it all for his sake and only said this to leave him no right to refuse. He remained silent, knitting his brows. He felt that he was so surrounded and involved in a mesh of falsity that it was hard to unravel anything.

Everything she did for him was entirely for her own sake, and she told him she was doing for herself what she actually was doing for herself, as if that was so incredible that he must understand the opposite.

At half-past eleven the celebrated specialist arrived. Again the sounding began and the significant conversations in his presence and in another room, about the kidneys and the appendix, and the questions and answers, with such an air of importance that again, instead of the real question of life and death which now alone confronted him, the question arose of the kidney and appendix which were not behaving as they ought to and would now be attacked by Michael Danilovich and the specialist and forced to amend their ways.

The celebrated specialist took leave of him with a serious though not hopeless look, and in reply to the timid question Ivan Ilych, with eyes glistening with fear and hope, put to him as to whether there was a chance of recovery, said that he could not vouch for it but there was a possibility. The look of hope with which Ivan Ilych watched the doctor out was so pathetic that Praskovya Fëdorovna, seeing it, even wept as she left the room to hand the doctor his fee.

The gleam of hope kindled by the doctor's encouragement did not last long. The same room, the same pictures, curtains, wall-paper, medicine bottles, were all there, and the same aching suffering body, and Ivan Ilych began to moan. They gave him a subcutaneous injection and he sank into oblivion.

It was twilight when he came to. They brought him his dinner and he swallowed some beef tea with difficulty, and then everything was the same again and night was coming on.

After dinner, at seven o'clock, Praskovya Fëdorovna came into the room in evening dress, her full bosom pushed up by her corset, and with traces of powder on her face. She had reminded him in the morning that they were going to the theatre. Sarah Bernhardt was visiting the town and they had a box, which he had insisted on their taking. Now he had forgotten about it and her toilet offended him, but he concealed his vexation when he remembered that he had himself insisted on their securing a box and going because it would be an instructive and aesthetic pleasure for the children.

Praskovya Fëdorovna came in, self-satisfied but yet with a rather guilty air. She sat down and asked how he was, but, as he saw, only for the sake of asking and not in order to learn about it, knowing that there was nothing to learn — and then went on to what she really wanted to say: that she would not on any account have gone but that the box had been taken and Helen and their daughter were going, as well as Petrishchev (the examining magistrate, their daughter's fiancé) and that it was out of the question to let them go alone; but that she would have much preferred to sit with him for a while; and he must be sure to follow the doctor's orders while she was away.

"Oh, and Fëdor Petrovich" (the fiancé) "would like to come in, may he? And Lisa?"

"All right."

Their daughter came in in full evening dress, her fresh young flesh exposed (making a show of that very flesh which in his own case caused so much suffering), strong, healthy, evidently in love, and impatient with illness, suffering, and death, because they interfered with her happiness.

Fëdor Petrovich came in too, in evening dress, his hair curled á la Capoul, a tight stiff collar round his long sinewy neck, an enormous white shirt-front and narrow black trousers tightly stretched over his strong thighs. He had one white glove tightly drawn on, and was holding his opera hat in his hand.

Following him the schoolboy crept in unnoticed, in a new uniform, poor little fellow, and wearing gloves. Terribly dark shadows showed under his eyes, the meaning of which Ivan Ilych knew well.

His son had always seemed pathetic to him, and now it was dreadful to see the boy's frightened look of pity. It seemed to Ivan Ilych that Vasya was the only one besides Gerasim who understood and pitied him.

They all sat down and again asked how he was. A silence followed. Lisa asked her mother about the opera-glasses, and there was an altercation between mother and daughter as to who had taken them and where they had been put. This occasioned some unpleasantness.

Fëdor Petrovich inquired of Ivan Ilych whether he had ever seen Sarah Bernhardt. Ivan Ilych did not at first catch the question, but then replied: "No, have you seen her before?"

"Yes, in Adrienne Lecouvreur."

Praskovya Fëdorovna mentioned some rôles in which Sarah Bernhardt was particularly good. Her daughter disagreed. Conversation sprang up as to the elegance and realism of her acting – the sort of conversation that is always repeated and is always the same.

In the midst of the conversation Fëdor Petrovich glanced at Ivan Ilych and became silent. The others also looked at him and grew silent. Ivan Ilych was staring with glittering eyes straight before him, evidently indignant with them. This had to be rectified, but it was impossible to do so. The silence had to be broken, but for a time no one dared to break it and they all became afraid that the conventional deception would suddenly become obvious and the truth become plain to all. Lisa was the first to pluck up courage and break that silence, but by trying to hide what everybody was feeling, she betrayed it.

"Well, if we are going it's time to start," she said, looking at her watch, a present from her father, and with a faint and significant smile at Fëdor Petrovich relating to something known only to them. She got up with a rustle of her dress.

They all rose, said good-night, and went away.

When they had gone it seemed to Ivan Ilych that he felt better; the falsity had gone with them. But the pain remained — that same pain and that same fear that made everything monotonously alike, nothing harder and nothing easier. Everything was worse.

Again minute followed minute and hour followed hour. Everything remained the same and there was no cessation. And the inevitable end of it all became more and more terrible.

"Yes, send Gerasim here," he replied to a question Peter asked.

His wife returned late at night. She came in on tiptoe, but he heard her, opened his eyes, and made haste to close them again. She wished to send Gerasim away and to sit with him herself, but he opened his eyes and said: "No, go away."

"Are you in great pain?"

"Always the same."

"Take some opium."

He agreed and took some. She went away.

Till about three in the morning he was in a state of stupefied misery. It seemed to him that he and his pain were being thrust into a narrow, deep black sack, but though they were pushed further and further in they could not be pushed to the bottom. And this, terrible enough in itself, was accompanied by suffering. He was frightened yet wanted to fall through the sack, he struggled but yet co-operated. And suddenly he broke through, fell, and regained consciousness. Gerasim was sitting at the foot of the bed dozing quietly and patiently, while he himself lay with his emaciated stockinged legs resting on Gerasim's shoulders; the same shaded candle was there and the same unceasing pain.

"Go away, Gerasim," he whispered.

"It's all right, sir. I'll stay a while."

"No. Go away."

He removed his legs from Gerasim's shoulders, turned sideways onto his arm, and felt sorry for himself. He only waited till Gerasim had gone into the next room and then restrained himself no longer but wept like a child. He wept on account of his helplessness, his terrible loneliness, the cruelty of man, the cruelty of God, and the absence of God.

"Why hast Thou done all this? Why hast Thou brought me here? Why, why dost Thou torment me so terribly?"

He did not expect an answer and yet wept because there was no answer and could be none. The pain again grew more acute, but he did not stir and did not call. He said to himself: "Go on! Strike me! But what is it for? What have I done to Thee? What is it for?"

Then he grew quiet and not only ceased weeping but even held his breath and became all attention. It was as though he were listening not to an audible voice but to the voice of his soul, to the current of thoughts arising within him.

"What is it you want?" was the first clear conception capable of expression in words, that he heard.

"What do you want? What do you want?" he repeated to himself.

"What do I want? To live and not to suffer," he answered.

And again he listened with such concentrated attention that even his pain did not distract him.

"To live? How?" asked his inner voice.

"Why, to live as I used to — well and pleasantly."

"As you lived before, well and pleasantly?" the voice repeated.

And in imagination he began to recall the best moments of his pleasant life. But strange to say none of those best moments of his pleasant life now seemed at all what they had then seemed — none of them except the first recollections of childhood. There, in childhood, there had been something really pleasant with which it would be possible to live if it could return. But the child who had experienced that happiness existed no longer, it was like a reminiscence of somebody else.

As soon as the period began which had produced the present Ivan Ilych, all that had then seemed joys now melted before his sight and turned into something trivial and often nasty.

And the further he departed from childhood and the nearer he came to the present the more worthless and doubtful were the joys. This began with the School of Law. A little that was really good was still found there — there was light-heartedness, friendship, and hope. But in the upper classes there had already been fewer of such good moments. Then during the first years of his official career, when he was in the service of the Governor, some pleasant moments again occurred: they were the memories of love for a woman. Then all became confused and there was still less of what was good; later on again there was still less that was good, and the further he went the less there was. His marriage, a mere accident, then the disenchantment that followed it, his wife's bad breath and the sensuality and hypocrisy: then that deadly official life and those preoccupations about money, a year of it, and two, and ten, and twenty, and always the same thing. And the longer it lasted the more deadly it became. "It is as if I had been going downhill while I imagined I was going up. And that is really what it was. I was going up in public opinion, but to the same extent life was ebbing away from me. And now it is all done and there is only death."

"Then what does it mean? Why? It can't be that life is so senseless and horrible. But if it really has been so horrible and senseless, why must I die and die in agony? There is something wrong!"

"Maybe I did not live as I ought to have done," it suddenly occurred to him. "But how could that be, when I did everything properly?" he replied, and immediately dismissed from his mind this, the sole solution of all the riddles of life and death, as something quite impossible.

"Then what do you want now? To live? Live how? Live as you lived in the law courts when the usher proclaimed 'The judge is coming!' The judge is coming, the judge!" he repeated to himself. "Here he is, the judge. But I am not guilty!" he exclaimed angrily. "What is it for?" And he ceased crying, but turning his face to the wall continued to ponder on the same question: Why, and for what purpose, is there all this horror? But however much he pondered he found no answer. And whenever the thought occurred to him, as it often did, that it all resulted from his not having lived as he ought to have done, he at once recalled the correctness of his whole life and dismissed so strange an idea.

Another fortnight passed. Ivan Ilych now no longer left his sofa. He would not lie in bed but lay on the sofa, facing the wall nearly all the time. He suffered ever the same unceasing agonies and in his loneliness pondered always on the same insoluble question: "What is this? Can it be that it is Death?" And the inner voice answered: "Yes, it is Death."

"Why these sufferings?" And the voice answered, "For no reason — they just are so." Beyond and besides this there was nothing.

From the very beginning of his illness, ever since he had first been to see the doctor, Ivan Ilych's life had been divided between two contrary and alternating moods: now it was despair and the expectation of this uncomprehended and terrible death, and now hope and an intently interested observation of the functioning of his organs. Now before his eyes there was only a kidney or an intestine that temporarily evaded its duty, and now only that incomprehensible and dreadful death from which it was impossible to escape.

These two states of mind had alternated from the very beginning of his illness, but the further it progressed the more doubtful and fantastic became the conception of the kidney, and the more real the sense of impending death.

He had but to call to mind what he had been three months before and what he was now, to call to mind with what regularity he had been going downhill, for every possibility of hope to be shattered.

Latterly during that loneliness in which he found himself as he lay facing the back of the sofa, a loneliness in the midst of a populous town and surrounded by numerous acquaintances and relations but that yet could not have been more complete anywhere — either at the bottom of the sea or under the earth — during that terrible loneliness Ivan Ilych had lived only in memories of the past. Pictures of his past rose before him one after another. They always began with what was nearest in time and then went back to what was most remote — to his childhood — and rested there. If he thought of the stewed prunes that had been offered him that day, his mind went back to the raw shrivelled French plums of his childhood, their peculiar flavour and the flow of saliva when he sucked their stones, and along with the memory of that taste came a whole series of memories of those days: his nurse, his brother, and their toys. "No, I mustn't think of that .... It is too painful," Ivan Ilych said to himself, and brought himself back to the present — to the button on the back of the sofa and the creases in its morocco. "Morocco is expensive, but it does not wear well: there had been a quarrel about it. It was a different kind of quarrel and a different kind of morocco that time when we tore father's portfolio and were punished, and mamma brought us some tarts . . . . " And again his thoughts dwelt on his childhood, and again it was painful and he tried to banish them and fix his mind on something else.

Then again together with that chain of memories another series passed through his mind — of how his illness had progressed and grown worse. There also the further back he looked the more life there had been. There had been more of what was good in life and more of life itself. The two merged together. "Just as the pain went on getting worse and worse, so my life grew worse and worse," he thought. "There is one bright spot there at the back, at the beginning of life, and afterwards all becomes blacker, and blacker and proceeds more and more rapidly — in inverse ratio to the square of the distance from death," thought Ivan Ilych. And the example of a stone falling downwards with increasing velocity entered his mind. Life, a series of increasing sufferings, flies further and further towards its end — the most terrible suffering. "I am flying . . . . " He shuddered, shifted himself, and tried to resist, but was already aware that resistance was impossible, and again with eyes weary of gazing but unable to cease seeing what was before them, he stared at the back of the sofa and waited — awaiting that dreadful fall and shock and destruction.

"Resistance is impossible!" he said to himself. "If I could only understand what it is all for! But that too is impossible. An explanation would be possible if it could be said that I have not lived as I ought to. But it is impossible to say that," and he remembered all the legality, correctitude, and propriety of his life. "That at any rate can certainly not be admitted," he thought, and his lips smiled ironically as if someone could see that smile and be taken in by it. "There is no explanation! Agony, death . . . . What for?"

Another two weeks went by in this way and during that fortnight an event occurred that Ivan Ilych and his wife had desired. Petrishchev formally proposed. It happened in the evening. The next day Praskovya Fëdorovna came into her husband's room considering how best to inform him of it, but that very night there had been a fresh change for the worse in his condition. She found him still lying on the sofa but in a different position. He lay on his back, groaning and staring fixedly straight in front of him.

She began to remind him of his medicines, but he turned his eyes towards her with such a look that she did not finish what she was saying; so great an animosity, to her in particular, did that look express.

"For Christ's sake let me die in peace!" he said.

She would have gone away, but just then their daughter came in and went up to say good morning. He looked at her as he had done at his wife, and in reply to her inquiry about his health said dryly that he would soon free them all of himself. They were both silent and after sitting with him for a while went away.

"Is it our fault?" Lisa said to her mother. "It's as if we were to blame! I am sorry for papa, but why should we be tortured?"

The doctor came at his usual time. Ivan Ilych answered "Yes" and "No," never taking his angry eyes from him, and at last said: "You know you can do nothing for me, so leave me alone."

"We can ease your sufferings."

"You can't even do that. Let me be."

The doctor went into the drawing-room and told Praskovya Fëdorovna that the case was very serious and that the only resource left was opium to allay her husband's sufferings, which must be terrible.

It was true, as the doctor said, that Ivan Ilych's physical sufferings were terrible, but worse than the physical sufferings were his mental sufferings, which were his chief torture.

His mental sufferings were due to the fact that that night, as he looked at Gerasim's sleepy, good-natured face with its prominent cheek-bones, the question suddenly occurred to him: "What if my whole life has really been wrong?"

It occurred to him that what had appeared perfectly impossible before, namely that he had not spent his life as he should have done, might after all be true. It occurred to him that his scarcely perceptible attempts to struggle against what was considered good by the most highly placed people, those scarcely noticeable impulses which he had immediately suppressed, might have been the real thing, and the rest false. And his professional duties and the whole arrangement of his life and of his family, and all his social and official interests, might all have been false. He tried to defend all those things to himself and suddenly felt the weakness of what he was defending. There was nothing to defend.

"But if that is so," he said to himself, "and I am leaving this life with the consciousness that I have lost all that was given me and it is impossible to rectify it — what then?"

He lay on his back and began to pass his life in review in quite a new way. In the morning when he saw first his footman, then his wife, then his daughter, and then the doctor, their every word and movement confirmed to him the awful truth that had been revealed to him during the night. In them he saw himself — all that for which he had lived — and saw clearly that it was not real at all, but a terrible and huge deception which had hidden both life and death. This consciousness intensified his physical suffering tenfold. He groaned and tossed about, and pulled at his clothing which choked and stifled him. And he hated them on that account.

He was given a large dose of opium and became unconscious, but at noon his sufferings began again. He drove everybody away and tossed from side to side.

His wife came to him and said:

"Jean, my dear, do this for me. It can't do any harm and often helps. Healthy people often do it."

He opened his eyes wide.

"What? Take communion? Why? It's unnecessary! However . . . "

She began to cry.

"Yes, do, my dear. I'll send for our priest. He is such a nice man."

"All right. Very well," he muttered.

When the priest came and heard his confession, Ivan Ilych was softened and seemed to feel a relief from his doubts and consequently from his sufferings, and for a moment there came a ray of hope. He again began to think of the vermiform appendix and the possibility of correcting it. He received the sacrament with tears in his eyes.

When they laid him down again afterwards he felt a moment's ease, and the hope that he might live awoke in him again. He began to think of the operation that had been suggested to him. "To live! I want to live!" he said to himself.

His wife came in to congratulate him after his communion, and when uttering the usual conventional words she added:

"You feel better, don't you?"

Without looking at her he said "Yes."

Her dress, her figure, the expression of her face, the tone of her voice, all revealed the same thing. "This is wrong, it is not as it should be. All you have lived for and still live for is falsehood and deception, hiding life and death from you." And as soon as he admitted that thought, his hatred and his agonizing physical suffering again sprang up, and with that suffering a consciousness of the unavoidable, approaching end. And to this was added a new sensation of grinding shooting pain and a feeling of suffocation.

The expression of his face when he uttered that "yes" was dreadful. Having uttered it, he looked her straight in the eyes, turned on his face with a rapidity extraordinary in his weak state and shouted:

"Go away! Go away and leave me alone!"

From that moment the screaming began that continued for three days, and was so terrible that one could not hear it through two closed doors without horror. At the moment he answered his wife he realized that he was lost, that there was no return, that the end had come, the very end, and his doubts were still unsolved and remained doubts.

"Oh! Oh! Oh!" he cried in various intonations. He had begun by screaming "I won't!" and continued screaming on the letter O.

For three whole days, during which time did not exist for him, he struggled in that black sack into which he was being thrust by an invisible, resistless force. He struggled as a man condemned to death struggles in the hands of the executioner, knowing that he cannot save himself. And every moment he felt that despite all his efforts he was drawing nearer and nearer to what terrified him. He felt that his agony was due to his being thrust into that black hole and still more to his not being able to get right into it. He was hindered from getting into it by his conviction that his life had been a good one. That very justification of his life held him fast and prevented his moving forward, and it caused him most torment of all.

Suddenly some force struck him in the chest and side, making it still harder to breathe, and he fell through the hole and there at the bottom was a light. What had happened to him was like the sensation one sometimes experiences in a railway carriage when one thinks one is going backwards while one is really going forwards and suddenly becomes aware of the real direction.

"Yes, it was all not the right thing," he said to himself, "but that's no matter. It can be done. But what is the right thing?" he asked himself, and suddenly grew quiet.

This occurred at the end of the third day, two hours before his death. Just then his schoolboy son had crept softly in and gone up to the bedside. The dying man was still screaming desperately and waving his arms. His hand fell on the boy's head, and the boy caught it, pressed it to his lips, and began to cry.

At that very moment Ivan Ilych fell through and caught sight of the light, and it was revealed to him that though his life had not been what it should have been, this could still be rectified. He asked himself, "What is the right thing?" and grew still, listening. Then he felt that someone was kissing his hand. He opened his eyes, looked at his son, and felt sorry for him. His wife came up to him and he glanced at her. She was gazing at him open-mouthed, with undried tears on her nose and cheek and a despairing look on her face. He felt sorry for her too.

"Yes, I am making them wretched," he thought. "They are sorry, but it will be better for them when I die." He wished to say this but had not the strength to utter it. "Besides, why speak? I must act," he thought. With a look at his wife he indicated his son and said: "Take him away . . . sorry for him . . . sorry for you too . . . " He tried to add, "forgive me," but said "forgo" and waved his hand, knowing that He whose understanding mattered would understand.

And suddenly it grew clear to him that what had been oppressing him and would not leave him was all dropping away at once from two sides, from ten sides, and from all sides. He was sorry for them, he must act so as not to hurt them: release them and free himself from these sufferings. "How good and how simple!" he thought. "And the pain?" he asked himself. "What has become of it? Where are you, pain?"

He turned his attention to it.

"Yes, here it is. Well, what of it? Let the pain be."

"And death . . .where is it?"

He sought his former accustomed fear of death and did not find it. "Where is it? What death?" There was no fear because there was no death.

In place of death there was light.

"So that's what it is!" he suddenly exclaimed aloud. "What joy!"

To him all this happened in a single instant, and the meaning of that instant did not change. For those present his agony continued for another two hours. Something rattled in his throat, his emaciated body twitched, then the gasping and rattle became less and less frequent.

"It is finished!" said someone near him.

He heard these words and repeated them in his soul.

"Death is finished," he said to himself. "It is no more!"

He drew in a breath, stopped in the midst of a sigh, stretched out, and died.

A Natural History of the Dead

Ernest Hemingway

AUTHOR: There is not a word of conversation in the chapter, Madame, yet we have reached the end. I'm very sorry.

OLD LADY: No sorrier than I am, sir.

AUTHOR: What would you like to have? More major truths about the passions of the race? A diatribe against venereal disease? A few bright thoughts on death and dissolution? Or would you care to hear the author's experience with a porcupine during his earliest years spent in Emmett and Charlevoix counties in the state of Michigan?

Please, sir, no more about animals to-day.

What do you say to one of those homilies on life and death that delight an author so to write?

I cannot truly say I want that either. Have you not something of a sort I've never read, amusing yet instructive? I do not feel my best to-day.

Madame, I have the very thing you need. It's not about wild animals nor bulls. It's written in popular style and is designed to be the Whittier's Snow Bound of our time and at the end it's simply full of conversation.

If it has conversation in it I would like to read it.

Do so then, it's called

A NATURAL HISTORY OF THE DEAD

OLD LADY: I don't care for the title.

AUTHOR: I didn't say you would. You may very well not like any of it. But here it is:

A NATURAL HISTORY OF THE DEAD

It has always seemed to me that the war has been omitted as a field for the observations of the naturalist. We have charming and sound accounts of the flora and fauna of Patagonia by the late W. H. Hudson; the Reverend Gilbert White has written most interestingly of the Hoopoe on its occasional and not at all common visits to Selborne and Bishop Stanley has given us a valuable, although popular, Familiar History of Birds. Can we not hope to furnish the reader with a few rational and interesting facts about the dead? I hope so.

When that persevering traveller, Mungo Park, was at one period of his course fainting in the vast wilderness of an African desert, naked and alone, considering his days as numbered and nothing appearing to remain for him to do but to lie down and die, a small moss-flower of extraordinary beauty caught his eye. "Though the whole plant," says he, "was no larger than one of my fingers, I could not contemplate the delicate conformation of its roots, leaves and capsules without admiration. Can that Being who planted, watered and brought to perfection, in this obscure part of the world, a thing which appears of so small importance, look with unconcern upon the situation and suffering of creatures formed after his own image? Surely not. Reflections like these would not allow me to despair; I started up and, disregarding both hunger and fatigue, travelled forward, assured that relief was at hand; and I was not disappointed."

With a disposition to wonder and adore in like manner, as Bishop Stanley says, can no branch of Natural History be studied without increasing that faith, love and hope which we also, every one of us, need in our journey through the wilderness of life? Let us therefore see what inspiration we may derive from the dead.

In war the dead are usually the male of the human species although this does not hold true with animals, and I have frequently seen dead mares among the horses. An interesting aspect of war, too, is that it is only there that the naturalist has an opportunity to observe the dead of mules. In twenty years of observation in civil life I had never seen a dead mule and had begun to entertain doubts as to whether these animals were really mortal. On rare occasions I had seen what I took to be dead mules, but on close approach these always proved to be living creatures who seemed to be dead through their quality of complete repose. But in war these animals succumb in much the same manner as the more common and less hardy horse.

OLD LADY: I thought you said it wasn't about animals.

AUTHOR: It won't be for long. Be patient, can't you? It's very hard to write like this.

Most of those mules that I saw dead were along mountain roads or lying at the foot of steep declivities whence they had been pushed to rid the road of their encumbrance. They seemed a fitting enough sight in the mountains where one was accustomed to their presence and looked less incongruous there than they did later, at Smyrna, where the Greeks broke the legs of all their baggage animals and pushed them off the quay into the shallow water to drown. The numbers of broken-legged mules and horses drowning in the shallow water called for a Goya to depict them. Although, speaking literally, one can hardly say that they called for a Goya since there has only been one Goya, long dead, and it is extremely doubtful if these animals, were they able to call, would call for pictorial representation of their plight but, more likely, would, if they were articulate, call for some one to alleviate their condition.

OLD LADY: You wrote about those mules before.

AUTHOR: I know it and I'm sorry. Stop interrupting. I won't write about them again. I promise.

Regarding the sex of the dead it is a fact that one becomes so accustomed to the sight of all the dead being men that the sight of a dead woman is quite shocking. I first saw inversion of the usual sex of the dead after the explosion of a munition factory which had been situated in the countryside near Milan, Italy. We drove to the scene of the disaster in trucks along poplar-shaded roads, bordered with ditches containing much minute animal life, which I could not clearly observe because of the great clouds of dust raised by the trucks. Arriving where the munition plant had been, some of us were put to patrolling about those large stocks of munitions which for some reason had not exploded, while others were put at extinguishing a fire which had gotten into the grass of an adjacent field, which task being concluded, we were ordered to search the immediate vicinity and surrounding fields for bodies. We found and carried to an improvised mortuary a good number of these and, I must admit, frankly, the shock it was to find that these dead were women rather than men. In those days women had not yet commenced to wear their hair cut short, as they did later for several years in Europe and America, and the most disturbing thing, perhaps because it was the most unaccustomed, was the presence and, even more disturbing, the occasional absence of this long hair. I remember that after we had searched quite thoroughly for the complete dead we collected fragments. Many of these were detached from a heavy, barbed-wire fence which had surrounded the position of the factory and from the still existent portions of which we picked many of these detached bits which illustrated only too well the tremendous energy of high explosive. Many fragments we found a considerable distance away in the fields, they being carried farther by their own weight. On our return to Milan I recall one or two of us discussing the occurrence and agreeing that the quality of unreality and the fact that there were no wounded did much to rob the disaster of a horror which might have been much greater. Also the fact that it had been so immediate and that the dead were in consequence still as little unpleasant as possible to carry and deal with made it quite removed from the usual battlefield experience. The pleasant, though dusty, ride through the beautiful Lombard countryside also was a compensation for the unpleasantness of the duty and on our return, while we exchanged impressions, we all agreed that it was indeed fortunate that the fire which broke out just before we arrived had been brought under control as rapidly as it had and before it had attained any of the seemingly huge stocks of unexploded munitions. We agreed too that the picking up of the fragments had been an extraordinary business; it being amazing that the human body should be blown into pieces which exploded along no anatomical lines, but rather divided as capriciously as the fragmentation in the burst of a high explosive shell.

OLD LADY: This is not amusing.

AUTHOR: Stop reading it then. Nobody makes you read it. But please stop interrupting.

A naturalist, to obtain accuracy of observation, may confine himself in his observations to one limited period and I will take first that following the Austrian offensive of June, 1918, in Italy as one in which the dead were present in their greatest numbers, a withdrawal having been forced and an advance later made to recover the ground lost so that the positions after the battle were the same as before except for the presence of the dead. Until the dead are buried they change somewhat in appearance each day. The color change in Caucasian races is from white to yellow, to yellow-green, to black. If left long enough in the heat the flesh comes to resemble coal-tar, especially where it has been broken or torn, and it has quite a visible tarlike iridescence. The dead grow larger each day until sometimes they become quite too big for their uniforms, filling these until they seem blown tight enough to burst. The individual members may increase in girth to an unbelievable extent and faces fill as taut and globular as balloons. The surprising thing, next to their progressive corpulence, is the amount of paper that is scattered about the dead. Their ultimate position, before there is any question of burial, depends on the location of the pockets in the uniform. In the Austrian army these pockets were in the back of the breeches and the dead, after a short time, all consequently lay on their faces, the two hip pockets pulled out and, scattered around them in the grass, all those papers their pockets had contained. The heat, the flies, the indicative positions of the bodies in the grass and the amount of paper scattered are the impressions one retains. The smell of a battlefield in hot weather one cannot recall. You can remember that there was such a smell, but nothing ever happens to you to bring it back. It is unlike the smell of a regiment, which may come to you suddenly while riding in the street car and you will look across and see the man who has brought it to you. But the other thing is gone as completely as when you have been in love; you remembers things that happened, but the sensation cannot be recalled.

OLD LADY: I like it whenever you write about love.

AUTHOR: Thank you, Madame.

One wonders what that persevering traveller, Mungo Park, would have seen on a battlefield in hot weather to restore his confidence. There were always poppies in the wheat in the end of June and in July, and the mulberry trees were in full leaf and one could see the heat waves rise from the barrels of the guns where the sun struck them through the screens of leaves; the earth was turned a bright yellow at the edge of holes where mustard gas shells had been and the average broken house is finer to see than one that has never been shelled, but few travellers would take a good full breath of that early summer air and have any such thoughts as Mungo Park about those formed in His own image.

The first thing that you found about the dead was that, hit badly enough, they died like animals. Some quickly from a little wound you would not think would kill a rabbit. They died from little wounds as rabbits die sometimes from three or four small grains of shot that hardly seem to break the skin. Others would die like cats, a skull broken in and iron in the brain, they lie alive two days like cats that crawl into the coal bin with a bullet in the brain and will not die until you cut their heads off. Maybe cats do not die then, they say they have nine lives, I do not know, but most men die like animals, not men. I'd never seen a natural death, so called, and so I blamed it on the war and like the persevering traveller, Mungo Park, knew that there was something else, that always absent something else, and then I saw one.

The only natural death I've ever seen, outside of loss of blood, which isn't bad, was death from Spanish influenza. In this you drown in mucus, choking, and how you know the patient's dead is; at the end he shits the bed full. So now I want to see the death of any self-called Humanist because a persevering traveller like Mungo Park or me lives on and maybe yet will live to see the actual death of members of this literary sect and watch the noble exits that they make. In my musings as a naturalist it has occurred to me that while decorum is an excellent thing some must be indecorous if the race is to be carried on since the position prescribed for procreation is indecorous, highly indecorous, and it occurred to me that perhaps that is what these people are, or were; the children of decorous cohabitation. But regardless of how they started I hope to see the finish of a few, and speculate how worms will try that long preserved sterility; with their quaint pamphlets gone to bust and into foot-notes all their lust.

OLD LADY: That's a very nice line about lust.

AUTHOR: I know it. It came from Andrew Marvell. I learned how to do that by reading T. S. Eliot.

OLD LADY: The Eliots were all old friends of our family. I believe they were in the lumber business.

AUTHOR: My uncle married a girl whose father was in the lumber business.

OLD LADY: How interesting.

While it is, perhaps, legitimate to deal with these self-designated citizens in a natural history of the dead, even though the designation may mean nothing by the time this work is published, yet it is unfair to the other dead, who were not dead in their youth of choice, who owned no magazines, many of whom had doubtless never even read a review, that one has seen in the hot weather with a half-pint of maggots working were their mouths have been. It was not always hot weather for the dead, much of the time it was the rain that washed them clean when they lay in it and made the earth soft when they were buried in it and sometimes then kept on until the earth was mud and washed them out and you had to bury them again. Or in the winter in the mountains you had to put them in the snow and when the snow melted in the spring some one else had to bury them. They had beautiful burying grounds in the mountains, war in the mountains is the most beautiful of all war, and in one of them, at a place called Pocol, they buried a general who was shot through the head by a sniper. This is where those writers are mistaken who write books called <u>Generals Die in Bed</u>, because this general died in a trench dug in snow, high in the mountains, wearing an Alpini hat with an eagle feather in it and a hole in front you couldn't put your little finger in and a hole in back you could put your fist in, if it were a small fist and you wanted to put it there, and much blood in the snow. He was a damned fine general, and so was General von Behr who commanded the Bavarian Alpenkorps troops at the battle of Caporetto and was killed in his staff car by the Italian rearguard as he drove into Udine ahead of his troops, and the titles of all such books should be <u>Generals Usually Die in Bed</u>, if we are to have any sort of accuracy in such things.

OLD LADY: When does the story start?

AUTHOR: Now, Madame, at once. You'll soon have it.

In the mountains too, sometimes, the snow fell on the dead outside the dressing station on the side that was protected by the mountain from any shelling. They carried them into a cave that had been dug into the mountainside before the earth froze. It was in this cave that a man whose head was broken as a flower-pot may be broken, although it was all held together by membranes and a skillfully applied bandage now soaked and hardened, with the structure of his brain disturbed by a piece of broken steel in it, lay a day, a night, and a day. The stretcher-bearers asked the doctors to go in and have a look at him. They saw him each time they made a trip and even when they did not look at him they heard him breathing. The doctor's eyes were red and the lids swollen, almost shut from tear gas. He looked at the man twice; once in daylight, once with a flashlight. That too would have made a good etching for Goya, the visit with the flashlight, I mean. After looking at him the second time the doctor believed the stretcher-bearers when they said the soldier was still alive.

"What do you want me to do about it?" he asked.

There was nothing they wanted done. But after a while they asked permission to carry him out and lay him with the badly wounded.

"No. No. No!" said the doctor who was busy. "What's the matter? Are you afraid of him?"

"We don't like to hear him in there with the dead."

"Don't listen to him. If you take him out of there you will have to carry him right back in."

"We wouldn't mind that, Captain Doctor."

"No," said the doctor. "No. Didn't you hear me say no?"

"Why don't you give him an overdose of morphine?" asked an artillery officer who was waiting to have a wound in his arm dressed.

"Do you think that is the only use I have for morphine? Would you like me to have to operate without morphine? You have a pistol, go out and shoot him yourself."

"He's been shot already," said the officer. "If some of you doctors were shot you'd be different."

"Thank you very much," said the doctor waving a forceps in the air. "Thank you a thousand times. What about these eyes?" He pointed the forceps at them. "How would you like these?"

"Tear gas. We call it lucky if it's tear gas."

"Because you leave the line," said the doctor. "Because you come running here with your tear gas to be evacuated. You rub onions in your eyes."

"You are beside yourself. I do not notice your insults. You are crazy."

The stretcher-bearers came in.

"Captain Doctor," one of them said.

"Get out of here!" said the doctor.

They went out.

"I will shoot the poor fellow," the artillery officer said. "I am a humane man. I will not let him suffer."

"Shoot him then," said the doctor. "Shoot him. Assume the responsibility. I will make a report. Wounded shot by lieutenant of artillery in first curing post. Shoot him. Go ahead shoot him."

"You are not a human being."

"My business is to care for the wounded, not to kill them. That is for gentlemen of the artillery."

"Why don't you care for him then?"

"I have done so. I have done all that can be done."

"Why don't you send him down on the cable railway?"

"Who are you to ask me questions? Are you my superior officer? Are you in command of this dressing post? Do me the courtesy to answer."

The lieutenant of artillery said nothing. The others in the room were all soldiers and there were no other officers present.

"Answer me," said the doctor holding a needle up in his forceps. "Give me a response."

"F——k yourself," said the artillery officer.

"So," said the doctor. "So, you said that. All right. All right. We shall see."

The lieutenant of artillery stood up and walked toward him.

"F——k yourself," he said. "F——k yourself. F——k your mother. F——k your sister. . . ."

The doctor tossed the saucer full of iodine in his face. As he came toward him, blinded, the lieutenant fumbled for his pistol. The doctor skipped quickly behind him, tripped him and, as he fell to the floor, kicked him several times and picked up the pistol in his rubber gloves. The lieutenant sat on the floor holding his good hand to his eyes.

"I'll kill you!" he said. "I'll kill you as soon as I can see."

"I am the boss," said the doctor. "All is forgiven since you know I am the boss. You cannot kill me because I have your pistol. Sergeant! Adjutant! Adjutant!"

"The adjutant is at the cable railway," said the sergeant.

"Wipe out this officer's eyes with alcohol and water. He has got iodine in them. Bring me the basin to wash my hands. I will take the officer next."

"You won't touch me."

"Hold him tight. He is a little delirious."

One of the stetcher-bearers came in.

"Captain Doctor."

"What do you want?"

"The man in the dead-house————"

"Get out of here."

"Is dead, Captain Doctor. I thought you would be glad to know."

"See, my poor lieutenant? We dispute about nothing. In time of war we dispute about nothing."

"F——k you," said the lieutenant of artillery. He still could not see. "You've blinded me."

"It is nothing," said the doctor. "Your eyes will be all right. It is nothing. A dispute about nothing."

"Ayee! Ayee! Ayee!" suddenly screamed the lieutenant. "You have blinded me! You have blinded me!"

"Hold him tight," said the doctor. "He is in much pain. Hold him very tight."

OLD LADY: Is that the end? I thought you said it was like John Greenleaf Whittier's Snow Bound.

Madame, I'm wrong again. We aim so high and yet we miss the target.

OLD LADY: You know I like you less and less the more I know you.

Madame, it is always a mistake to know an author.

The Arithmetic
of Epidemics

The plague epidemic of 1635 began in London with the deaths of two boarders in a West End rooming house; Defoe estimated that 100,000 persons died before it ran its course. The bills of mortality carefully recorded the deaths per week as they climbed from the tens and twenties into the hundreds and thousands.

Defoe compared (in a chapter not reproduced here) the spread of the plague to that of a fire. An isolated house at the edge of a village may be destroyed by fire without harming any others; in the same manner, an isolated person (or family) may succumb to the plague without causing an epidemic. On the other hand, a house that burns within a city may easily set fire to its neighbors, and they to theirs, so that an entire city may be destroyed. The plague, too, within a city's congestion may spread from person to person until all have been taken ill.

The plague has not struck Europe since 1789 nor in recent times have typhus, smallpox, or other diseases decimated entire populations as they did so often in the past. Epidemics, however, do occur. The flu has struck periodically since the great epidemic of 1918; the Asian strain in 1957-8 and the Hong Kong strain in 1968-9 swept over the entire world.

To prepare and distribute vaccines and other medicines, public health officials must be able to predict the likely course of an epidemic disease. It is not enough to say that a disease may spread like wild fire; those with responsibility for a nation's health must make more precise predictions. Nor is it enough that in retrospect we can describe (as Camus does in his book, *The Plague*) how schools, churches, and other meeting halls were commandeered to serve as hospitals; the number of additional beds to care for the sick must be estimated in advance so that they will be ready when needed. We must have, therefore, an arithmetic of epidemics in order to make accurate predictions.

In the following paragraphs I shall develop a simple series of calculations, not so that each of us may join the Public Health Service, but rather so that we can understand a little better how complex biological problems are subjected to mathematical analyses.

To start the calculations, we might compare an epidemic with a chain reaction of the sort that leads to atomic explosions. One person, for example, might infect two others, these two then infect four, the four infect eight, and so on. *Life* Magazine once had a remarkable series of photographs illustrating this type of chain reaction. The entire floor of a large gymnasium was solidly covered with mouse traps, each one with two ping-pong balls resting on its set spring. A single ball was then tossed into the room. Within seconds the place was a bedlam. The air was filled with a myriad of bounding balls, each of which when it landed sent up two others to take its place.

The mathematical description of an epidemic is not difficult. The number of newly infected persons at a given time (N_t, where t represents time) equals 2^t. At times 0, 1, 2, 3, and so on, N_t equals 2^0 (=1), 2^1 (=2), 2^2 (=4), 2^3 (=8), and so on. The choice of 2 as the number of healthy persons a sick one might infect is arbitrary. If we let i represent the average number of persons contaminated by a sick individual, the equation would be written

$$N_t = i^t$$

The unit of time in these calculations would depend upon the disease. It represents the interval from the time of exposure to the disease to the time when the exposed person becomes infectious, that is, the incubation time for the disease.

The equation we have developed here is really a simpleminded one. No allowance has been made for a decrease of susceptible persons in the population; as an epidemic grows, unaffected persons become rarer and rarer. The equation, $N_t = i^t$, is fine for the first three or four cycles of infection when the total number of infected persons (past and present) is still small compared to the total number of persons in the entire population. The total number of affected persons grows rapidly however. It took only seconds, remember, for the gymnasium to teem with flying ping-pong balls and exploding mouse traps, and it takes only milliseconds to develop an atomic explosion.

How can the simple calculations given above be altered to compensate for a population of a limited number of persons? What complication can we introduce into the arithmetic to make the calculations more realistic? We must remember that the total number (N_{total}) of persons infected since the onset of the epidemic cannot possibly exceed the number (P) of persons in the population. As N_{total} approaches P in size, the number of persons infected by a sick individual (i) must become smaller and smaller. When there are no more susceptible persons (that is, when $N_{total} = P$), i will be reduced to zero.

These adjustments can be incorporated into our calculations quite easily. First, recall that if

$$N_t = i^t,$$

then

$$N_{t+1} = iN_t.$$

That is, the number of new cases of the disease equals i times the number of ill persons in the previous cycle of infection. Now, to allow for a decline in the number of new cases as the epidemic rages through the population, we merely introduce a new term (one of several that are able to accomplish our goal)

$$N_{t+1} = iN_t\left(1 - \frac{N_{\text{total}}}{P}\right)$$

When the epidemic is young, N_{total} is small and so the equation is virtually identical to

$$N_{t+1} = iN_t$$

On the other hand, as $(N_{\text{total}})/P$ approaches 1.00, the term in parenthesis approaches zero and the number of new cases approaches zero as well.

From a knowledge of the disease that threatens to spread through a population, its method of transmission, its incubation period, and the degree of congestion of the threatened population, public health officials can estimate t and i. They can then calculate the number of new cases expected during successive time intervals. The course the epidemic is expected to take can be plotted before it has actually occurred. If i cannot be reduced by early diagnosis and isolation and if N_{total} cannot be enlarged artificially by immunization through vaccination, at least beds can be prepared in which the ill can be cared for.

A numerical example illustrating the course of an imaginary epidemic in a population of 100,000 persons has been assembled in the accompanying table and graph. In this example, t (the incubation time of the disease) has been set for convenience at one-fourth of a month or about eight days. The number of persons that each sick one infects, at least during the early days of the epidemic, has been set at two ($i = 2$). A number of facts can be seen in the graph. The peak of the epidemic occurs near the first of January even though the first person became ill early in September. The number of newly infected persons drops off rapidly after the peak of the epidemic has passed; persons who have been missed by the disease are rare at this time. Some persons avoid the disease altogether; 15,000 of the 100,000 persons never become exposed under the conditions used in this example.

The course of this artificial epidemic resembles reasonably well that of the black plague described by Defoe except that the example is based on a nonfatal infection. If any person within this population of 100,000 took upon himself, as Defoe's narrator did, the task of describing the course of the epidemic, his account would be remarkably similar to *A Journal of the Plague Year*. Furthermore, he would have about one chance in seven of observing the entire course of the epidemic without being infected with the disease. Presumably Defoe's narrator and his friend Dr. Heath managed to escape infection during London's plague epidemic for similar accidental reasons.

Scientists are accused of speaking and writing a foreign tongue, a language of tables, charts, and mathematical symbols. To the untrained, the language of mathematics may be difficult but it need not be mysterious. Mathematical equations are used in science because the logical consequences that follow from an initial statement can be arrived at easily even though numerous steps intervene between start and finish. An audience can follow a blackboardful of computations, but an audience is generally lost (and, therefore, asleep) if it has to follow more than three or four steps of an imprecise oral argument.

Almost always biological phenomena must be simplified in order to make them conform to convenient mathematical manipulations. The practitioner must know the restraints these simplifications place upon him; he must know when they grossly distort his conclusions. It is better by far, however, to obtain close approximations by the proper use of simplified calculations than to wring one's hands over an otherwise insoluble predicament. Better a reasonably accurate curve upon which to base a judgment than to form no judgment. Better an act based on reason even though it errs than an irrational act that proves to be correct by accident alone.

Time	New Cases	Total Cases (new + old)
0	1	1
1	2	3
2	4	7
3	8	15
4	16	31
5	32	63
6	64	127
7	128	255
8	254	509
9	507	1,016
10	1,004	2,020
11	1,966	3,986
12	3,775	7,761
13	6,966	14,727
14	11,879	26,606
15	17,437	44,043
16	19,515	63,558
17	14,223	77,781
18	6,321	84,102
19	2,010	86,112
20	558	86,670
21	149	86,819
22	39	86,858
23	10	86,868
24	3	86,871
25	1	86,872

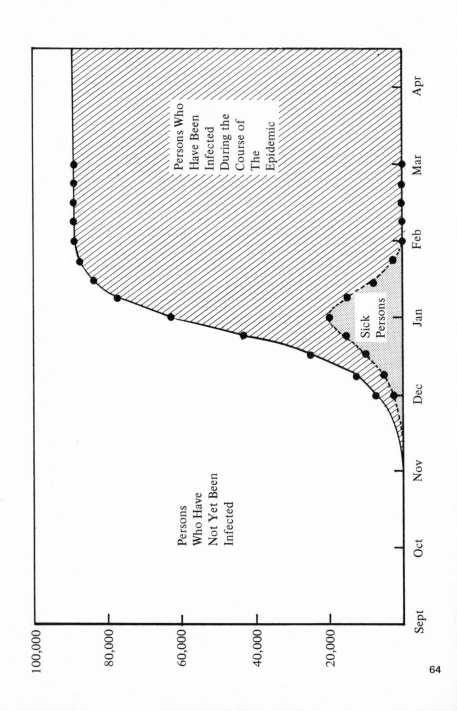

Epidemics
and Epidemic Diseases

Man's diseases are legion. Some await his birth and, sooner or later, attack him at home. Others lurk elsewhere and afflict him when he goes abroad for work or pleasure; an intestinal disorder known as "tourista" is well known to many Americans who have taken 15- and 21-day tours abroad. Still other diseases arrive from elsewhere; their progress can be measured as they sweep across cities as did the plague in London or as they hop from continent to continent as does the Asian flu in today's jet age.

Epidemic diseases can be classified in many ways: by causative organisms, for example, or by severity, or by symptoms on which diagnoses are made. Or we can classify them according to the "home" of both the disease and the victim: both residing in one place, the victim invading the home of the disease, or the disease invading the home of the victim. What an awkward classification! And yet it offers a rational basis for composing a rambling account of diseases, and so I shall use it in this essay.

Two diseases illustrate the ones that catch us at home — the common cold and poliomyelitis. Cold viruses are exceedingly infective. To go about one's daily activities in the office, store, post office, school, or church and to avoid being exposed to infectious viral particles expelled by the coughs and sneezes of others is virtually impossible. Immunity to a particular virus develops as a consequence of infection. And so, the five- or six-member teams that operate remote Arctic weather stations spend long periods free of cold symptoms; colds strike these persons shortly after the supply ship calls to leave mail and other material. Isolated villages also can be free of coughs and sneezes for long periods after all inhabitants have been exposed to and have developed immunity to the local viral strains. At a particular scientific community on Long Island, the third or fourth day of each annual ten-day international symposium brings on bleary eyes, runny noses, and hacking coughs as scientists from all continents exchange exotic cold viruses with one another and with their local hosts.

The common cold is a nuisance, not a threat. The few days inactivity that it enforces upon most of us may be a blessing in disguise; school children, at least, seem to enjoy the opportunity to skip classes. Polio, on the other hand, is an entirely different matter. Here is a disease that cripples and kills, one that often leaves its victim a helpless invalid. In seeming contradiction to logic, polio is a disease of advanced societies, a disease that accompanies improved sanitation and energetic health measures. Extraordinary efforts were expended in developing against this disease the vaccines that we now use routinely.

To understand polio is to understand many other diseases. Polio, like the common cold, is a viral disease that coexists with mankind. It is a disease that, if contracted early in life (in the first three years) is generally harmless; presumably at this age it passes as an ordinary cold with ordinary cold symptoms. Once a child has had polio, immunity exists for life.

Contracted late in life, polio is a serious disease; the older the victim; the more likely that death or paralysis will develop. The paradoxical correlation between polio and advanced public health measures lies in the "successful" passage of many children without exposure to the polio virus through the "safe" age, the age during which polio is a relatively harmless ailment, to a "nonsafe" age extending from preschool years onward, during which polio, if contracted, can lead to paralysis or death. Under primitive sanitary conditions, virtually all babies were inadvertently exposed to polio. As sanitation improved, the chances for exposure during the safe years dropped substantially; nevertheless, the chance for exposure during one's lifetime remained high. And so, the frequency of paralytic cases increased. Here, therefore, is a disease for which halfway preventive measures are worse than none at all. If some persons are going to be infected at all times, then it is best that all persons be infected at an age when the disease symptoms are mild. If substantial numbers are going to miss infection during the safe years, then it is necessary to completely eradicate the disease or to confer artificial immunity on all. The Salk vaccine and the Sabin oral vaccine are used for the latter purpose. Persons forty years of age or more should not take the Sabin vaccine however; should the live virus of this vaccine accidently lead to polio, the disease would be severe at this advanced age.

Diseases that lurk at home and infect native populations can infect strangers as well. Each of us has his intestinal flora consisting of bacteria, yeasts, fungi, and higher organisms. Do not underestimate these organisms; some of them (such as the common colon bacillus) are essential for our survival. Babies are born without a supply of these commensal organisms; they are picked up orally while nursing or while playing. At times they are acquired by purchase; pharmacists sell live *Lactobacillus acidophilous,* a bacterium that thrives well in human intestines, the bacterium that is used in manufacturing yogurt and cultured buttermilk.

The discomfort suffered by American tourists who go "native" in European or Latin American cities and villages is well known; flight bags carried by travelers contain paregoric and Entero-Vioform nearly as often as toothpaste and hand cream. American tourists do not always appreciate that Europeans and others who come to the United States and go "native" in our small towns (exchange professors in college towns, for example) suffer from the very same intestinal illnesses. Each of us is accustomed to the flora he has developed in his own body over many years. This rapport with our unseen friends is emphasized by the intestinal upset that frequently follows the use of antibiotics; the selective action of these medicines on the floral content of our large intestine can lead to novel proportions (an excess of yeast cells primarily) that cause at times acute discomfort.

Yellow fever, a disease of the tropics, has struck down large numbers of alien men who invaded its natural territories. These victims include the workers who strove to build the Panama Canal and the French soldiers who attempted to hold Haiti.

Yellow fever is transmitted by a species of mosquito, *Aedes aegypti*. The geographical range of the disease has at times been increased by the introduction of the yellow fever virus into an area where the mosquito had previously been free of the virus. South America, in fact, is a secondary focus of the disease, which is African in origin. The geographical range of the disease also has been altered drastically by the control of the carrier mosquito. Both Boston and Philadelphia have suffered epidemics in the past although there has been no urban outbreak within the United States since the early 1930's.

We have stressed the vulnerability of aliens to yellow fever. What about natives? Again, as in the case of polio, it appears that during early childhood yellow fever is not a severe disease so that adult natives carry an immunity acquired safely at an early age.

The last category of diseases to be discussed here are those such as plague, typhus, smallpox, and others that are introduced occasionally into a population where they are not ordinarily found and then roll through it with devastating results. Such epidemics swept over Europe in waves during past centuries. The plague came in the 9th century, then leprosy, and plague again (as the Black Death) in the mid-1300's. Sweating sickness struck England every ten or twenty years from 1485 until 1550. Syphilis ran rampant in Europe during the 1500's – not syphilis as we know it today but one that produced massive running sores extending from head to foot. Smallpox, tuberculosis, scarlet fever, diphtheria, and measles have all held sway in turn since the 18th century. Influenza has been the epidemic disease in the 20th century; cholera, since 1961.

Diseases that strike mankind with devastating invasive epidemics share certain features. Each has a reservoir from which periodically it emerges anew. For each disease, its characteristic symptoms ameliorate once an epidemic has run its course. The amelioration of symptoms deserves its own essay; here I shall discuss the reservoir of epidemic diseases.

What are the reservoirs of the diseases that cause severe epidemics? They are the populations of animal species for which each disease is more or less chronic and not particularly severe – perhaps resembling the common cold or the "cold sore" virus (herpes simplex) of man. The plague can be found in ground squirrels, prairie dogs, and marmots. The plague bacterium is transmitted from animal to animal by fleas. Before an epidemic of the plague hits man, it hits the rats that live with man. And as the bacterium is transmitted from ground squirrels and other rodents to rats, so it is transmitted from rats to man.

Typhus fevers, of which there are many sorts, are also carried by various rodents including rats and mice. In this case, the body louse (also a victim of the disease) spreads the virus from person to person in epidemic outbursts. Influenza, of which the Asian and Hong Kong flus have been recent examples, apparently has its origin in certain populations of the domestic pig; whether there is still another reservoir behind the pigs is unknown.

A frequently observed characteristic of epidemics is the mounting severity of the disease during the first weeks following its initial outbreak. According to some accounts, the increased severity results from the passage of the disease organism through susceptible individuals. This statement does not adequately explain the observation. A better understanding of the increased severity comes from the realization that (1) disease organisms differ among themselves (much like people) and (2) at the onset of an epidemic the disease has only recently found its way into man from its natural host animal. Among the individual bacteria or virus particles are some that can reproduce faster in man than can others. The increasing severity of the disease early in the epidemic reflects a turnover in the population of microorganisms with a loss of those varieties that are ill-adapted for life in man and an enormous increase in those that reproduce most rapidly. As a rule, the more rapidly the symptoms of a disease set in, the severer the disease and the less likely the victim is to recover.

The differential reproduction of various strains of microorganisms when these are first introduced into man is an illustration of *natural selection*. Differential reproduction of members of a variable population where the variation has a heritable basis (that is, like begets like) leads to alterations in the overall composition of the population. It may seem strange that natural selection can lead to the increased severity of a disease and, as we shall find out directly, to its amelioration as well. That, however, is the case and we shall learn about it in a later essay.

The Biology
of Some Diagnostic Symptoms

In Thasos, about the autumnal equinox, and under the Pleiades, the rains were abundant, constant, and soft, with southerly winds; the winter southerly, the northerly winds faint, droughts; on the whole, the winter having the character of spring. The spring was southerly, cool, rains small in quantity. Summer, for the most part, cloudy, no rain, the etesian winds, rare and small, blew in an irregular manner. The whole constitution of the season being thus inclined to the southerly, and with droughts early in the spring, from the preceding opposite and northerly state, ardent fevers occurred in a few instances, and these were very mild, being rarely attended with hemorrhage, and never proving fatal. Swellings appeared about the ears, in many on either side, and in the greatest number on both sides, being unaccompanied by fever so as not to confine the patient to bed; in all cases they disappeared without giving trouble, neither did any of them come to suppuration, as is common in swellings from other causes. They were of a lax, large, diffused character, without inflammation or pain, and they went away without any critical sign. They seized children, adults, and mostly those who were engaged in the exercises of the palestra and gymnasium, but seldom attacked women. Many had dry coughs without expectoration, and accompanied with hoarseness of voice. In some instances earlier, and in others later, inflammations with pain seized sometimes one testicle, and sometimes both; some of these cases were accompanied with fever and some not; the greater part of these were attended with much suffering. In other respects they were free of disease, so as not to require medical assistance.*

Thasos had a good winter! That was twenty-five centuries ago, of course. Hippocrates describes in the paragraph above not only the pleasant climate of Thasos but also the symptoms of mumps, easily recognizable, despite the great span of years, by those who have had the illness. Indeed, it reminds me of my own case when I was a high school senior and was quarantined during graduation week. The swellings, lax and painless, resulted in a superficial facial resemblance to Herbert Hoover (at that time, a recent president). They were "unaccompanied by fever . . . and disappeared without giving trouble." In this instance the disease did not "go down," the vernacular description for testicular involvement.

Hippocrates was an expert at diagnoses. "It appears to me," he wrote, "a most excellent thing for the physician to cultivate Prognosis; for by foreseeing and foretelling, in the presence of the sick, the present, the past, and the future, and explaining the omissions which patients have been guilty of, he will be the

*Reprinted from *Hippocrates: Ancient Medicine and Other Treatises* by permission of the Great Books Foundation. Copyright © by the Great Books Foundation.

more readily believed to be acquainted with the circumstances of the sick; so that men will have confidence to intrust themselves to such a physician." He then went on to describe facial appearances foretelling death that even now are known as *facies Hippocratica*. Similarly, the reflex movements of the hands — the movements of gathering bits of straw or picking the nap of the blanket — described as bad and deadly by Hippocrates are equally so today.

The diagnosis of a disease from gross bodily symptoms is not always an easy task. In the modern hospital, the visual and probing examination by the doctor is supplemented by a battery of appropriate physiological and biochemical analyses together with electronic tests. Indeed, if an infectious disease is involved, the studies usually include numerous tests to determine the resistance of the disease organism to various antibiotics. As little as possible of the prescribed therapy is left to chance.

The agenda of a routine office call today is very much as visits to doctors have been for centuries nevertheless. The urine is examined; the chest is listened to and thumped; reflexes are noted; the eyes, ears, nose, and throat are examined; the jowls are prodded. Furthermore, questions are asked because, in the final analysis, we are our own keepers. We live with ourselves and, for the most part, are responsible for noting occurrences of unexpected pains, sweating, tensions, or pressures. A fullness here, an ache there — these are our problems, not our neighbors. Even children become adept at foretelling which of their ailments will be minor and which threaten to be severe, which bouts of nausea will pass and which will lead to vomiting.

A short essay on the biology of diagnostic symptoms cannot touch on the great number of signs and symptoms looked for by doctors or on the even greater number of combinations in which these can occur. It cannot have as its purpose the training of medical experts. There is an intriguing constancy about the symptoms of man's ills, however, that suggest ancient and fundamental biological reactions to disease. Perhaps this should be called mis-biology of man. That the fall and winter in Thasos were mild and pleasant and that an epidemic of mumps hit the schoolchildren and adults of the city are interesting bits of history. For myself, I find it even more intriguing that I can distinguish on the basis of ancient writings between the mumps of one season and the recurrent fevers of malaria of another. What are the symptoms that make this possible? What are these swellings, these eruptions, these fevers?

The temperature control of the human body — man's thermostat — lies in his brain. Two centers of the brain exert an influence on the immediate temperature-regulating mechanisms of the body. These mechanisms are the dilation or constriction of surface blood vessels, sweating, shivering, and the liberation of adrenaline, a hormonal stimulant of heat production (increased oxidation).

Foreign proteins or even hemoglobin released from accidentally fragmented red blood cells stimulate the control centers to raise the body's temperature. The onset may consist of the constriction of surface blood vessels, a reduction in the volume and an increase in the concentration of the blood, and an increased production of heat.

The contraction of surface blood vessels allows the skin to cool rapidly and, since the body senses its temperature by surface receptors rather than interior ones, muscles are stimulated to contract rapidly and involuntarily; the patient shivers. The feverish person shakes with a chill, but shivering itself is a heat-producing reaction. It is an especially valuable one in case of exposure to freezing conditions; in this case shivering tends to compensate for the loss of heat. In feverish persons, however, shivering adds still more heat to a body that is already producing more heat than is normal and losing less by surface radiation. As a result, the body temperature rises until a new and higher equilibrium temperature is reached. Eventually the surface blood vessels may relax to bring on the flush commonly seen in feverish patients.

Certain diseases cause relapsing fevers; between bouts of high temperature the body temperature reverts to normal. Occasionally the pattern of recurrence is constant enough to be diagnostic in itself. Certain malarial parasites require 72 hours to mature in red blood cells. At that time the new crop of organisms is released by the rupture of the cells in which they developed; the newly emerged parasites immediately infect new cells, thus starting a new generation. The cyclic rupturing of red blood cells every 72 hours and the release of free hemoglobin in the blood stream results in a fever every three days. Other strains of malarial organisms require only 48 hours to mature; these strains result in fevers that recur every second day. Persons within which two or more "crops" of malarial parasites develop may have fevers that come and go every day. Physicians familiar with malaria can recognize the symptoms described in biblical and other ancient writings.

Plague sufferers have intense fevers because of the bacterial infection. The term "bubonic" refers to another symptom; it refers to swellings or buboes in the groin. These are swollen lymph nodes, nodes of tissue that are part of a circulatory system that is far less familiar than that consisting of the better-known arteries and veins.

Lymph is tissue fluid. The blood, having left the heart by way of the aorta, goes to all parts of the body. Large arteries branch from the aorta and these subdivide in turn until, finally, the arterial blood arrives at the capillaries, the tiny tubules that ramify and then reunite to form small veins. The veins anastomose and grow in size until the largest of them reach the heart to complete the bodily (as opposed to the pulmonary) circulatory cycle.

Lymph is fluid that has passed from the bloodstream into the tissues. The walls of the capillaries are permeable and a dilute serum of water and proteins diffuses from them into nearby tissue spaces. Some of this liquid diffuses back into the capillaries near their venous ends. The rest of it, however, is gathered into tubules that lead into larger ones to form the lymphatic system, a system of vessels that carries lymph (the dilute serum minus red blood cells) back to one of the large veins. Two large lymphatic ducts join this vein just before it enters the heart. And so there is a circulation of fluid from the peripheral capillaries and surrounding tissue to lymphatic vessels that eventually lead back to the main circulatory system once more.

At strategic places (groin, neck, back of the head, behind the knees, at the elbow, in the armpits, as well as numerous places internally) nodules straddle lymphatic vessels in such a manner that the lymph returning from the periphery must pass through a nodule (lymph node) before continuing to the heart. These lymph nodes have two functions. They manufacture one type of white blood cells (lymphocytes) that are essential in combating disease. Secondly, they serve as filters for the lymph that passes through them. If small foreign particles (soot for example) are injected into tissues, they will not enter the blood circulatory system; instead, such particles are carried in the lymph to an interceding node where they are filtered out. Bacteria are filtered from the circulating lymph too. In one experiment a solution containing 250 million streptococci per cubic centimeter was led into a dog's lymph duct for more than an hour, but the material emerging from the other side of the lymph node was sterile. The lymph node is able to protect an animal against a general blood "poisoning" (septicemia) despite a massive provocation of this sort.

Lymph nodes also collect the plague bacillus and enlarge to form buboes in plague victims. Similar swellings form from almost any infection. Ear infections can cause enlarged glands in the neck. An infected finger can cause swellings in the arm pit. In filtering out the bacteria, the lymph node activates the antibody-forming machinery of the body. The plague bacillus, however, generally overwhelms the defenses of the body; that is why, of course, it is such a severe disease. The lymph nodes enlarge and sometimes suppurate externally. In a high proportion of cases, they eventually fail to retain the bacteria, which then enter the bloodstream where they cause a general fatal infection.

Tonsils are lymphatic tissue but are not lymph nodes. They, too, serve as centers for the production of lymphocytes and presumably serve to filter out many bacteria that have entered the body through the nose and throat. Tonsils are subject to infection, however, and if these persist they render the tonsils a greater liability than an asset. Consequently, tonsillectomy is a common operation for growing children. Common or not, a tonsillectomy is a major operation and one that should not be undertaken except after a careful weighing of all arguments, pro and con.

Fevers and swellings are but two of the signs that reveal the presence of disease. Eruptions of the skin – the red spots pictured in the Sunday comics and TV cartoons – are another. These are the pocks of chicken pox and smallpox, for example, and the rash of measles. In these three cases, of which each is a viral disease, the eruptions reveal the extent to which the disease involves the patient.

These small eruptions are formed by the breakdown of virus-infected cells with a release of their viral content. A microscopic speck of the exudate of one such pustule would serve to inoculate a suitable host such as a susceptible person, a developing chick embryo, or the eye of a rabbit. Before the disease abates and resistance has been developed, the breakdown of infected cells and the release of viral particles capable of invading new cells has occurred throughout the body. The remarkable feature of these general infections, I suppose, is that defenses can be developed in time to halt their spread into areas where the infection would prove to be fatal. But that, indeed, is the case; only rarely do we encounter a complication, such as meningitis after chicken pox, for these common childhood diseases. Complications are so rare that many doctors are unfamiliar with their concomitant symptoms.

In anticipation of the next essay, I shall conclude by pointing out that the symptoms of some diseases have changed with time. The preface of this book cited a description of syphilis given by Voltaire in *Candide*; no longer are syphilitics covered by sores nor are their tissues devoured by the disease. Syphilis is still a serious and deadly disease to be sure; nevertheless, the raw violence of the early symptoms has abated over a period of several centuries.

Scarlet fever, even before the discovery of sulfa drugs and antibiotics, had undergone a similar diminution in severity. In the 1920's it was still a dreaded childhood disease. Its transmission was only poorly understood and precautions frequently included burning of clothing and books belonging to the patient. During the 1930's the symptoms had ameliorated to the extent that a new name, scarletina, was used more and more often. The causative organism is now know to be streptococcus and the scarlet of scarlet fever is recognized as an allergic reaction. The streptococci are so susceptible to penicillin treatment that the nature of the symptoms of scarlet fever is today an almost academic question.

On Natural Selection

Individual members of any population differ from one another. To some extent — a great extent in the case of some differences, a slight extent in the case of others — these differences are passed on from generation to generation. These differences are responsible at times for the failure of some individuals to leave offspring or, conversely, for the opportunity to leave an exceptionally large number. Consequently, the nature of the population changes with time; the array of characteristics present in the population in one generation is not identical to the array that existed in earlier generations nor that will exist in future ones. This is natural selection.

Natural selection is a consequence, not a cause. It is not a pagan deity endowed with a mystical ability to create species and comparable to Michelangelo's powerful Jehovah. If forms differ in the numbers of offspring they leave and if this difference has a genetic (or heritable) basis, then the nature of the collection of individuals composing the population must change with passing generations.

Consider a population of men. Many things are involved in determining the number of children a couple may have. Many children may be produced. Fewer may be produced but, because of greater parental care, a larger number may succeed in growing up. Persons capable of having children may refrain from reproducing for any one of a number of reasons. All of these possibilities are involved in altering the nature of human populations. Once a change has occurred, persons may look about and say, "See what natural selection has done."

A simple illustration of natural selection is found in bacterial populations. Imagine a city of some 100,000 persons but picture these people not as human beings but as vessels in and on which bacteria grow; the bacterium may be staphylococcus, the common cause of boils, carbuncles, and sore throats. These bacteria are spread both by contact and airborne particles. The spread from person to person is efficient enough so that we can think of a bacterial population that includes all the staphylococci of the city, a population that is housed in and continually being exchanged between 100,000 human carriers.

Periodically in the bacterial population a rare mutant type appears, an individual capable of making a virtually useless protein, an enzyme that can destroy penicillin. The word "useless" needs prompt attention. We are discussing here a population in the mid-1930's, a preantibiotic age. At that time the only penicillin in the world was that which oozed from the common green mold,

Penicillium. The staphylococcus population we are discussing had very little direct contact with this mold.

Each rare mutant individual expends a portion of its energy and of its material wealth in synthesizing a protein molecule for which it has no use. The extra drain on its body machinery delays the rate at which it can prepare for "division," and so its rate of reproduction is less than that of its normal neighbors.

The entire population of staphylococci in the city is more or less constant in size; thus, during each bacterial generation, an average of one new bacterium is made for each one already existing. Since the division of a bacterium gives rise to two daughter cells, some nondividing cells must die. The process of the replacement of one bacterial generation by the next can be pictured as follows.

Two generations of a bacterial population. Some bacteria die ● , others divide ⚇ , still others do neither ⚈ .

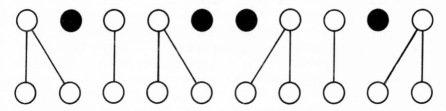

Within a single generation some bacteria die without dividing, some live but fail to divide, still others divide; at equilibrium the gains and losses are equal, and so the population remains about the same size.

The mutant type cannot maintain its own replacement on a scale comparable to that of the others; in each case, therefore, the mutant individual persists for just a relatively short time before it or its immediate descendants are lost.

In mathematical terms we say that the mutant arises anew by chance in a fraction, u, of all bacterial cells during each bacterial generation. In addition, we say that instead of each cell leaving a average of *one* descendant as in the diagram, mutant cells leave only $1 - s$ descendants (all mutant of course). Thus, mutant cells accumulate to a for-the-moment unspecified frequency, x, such that the loss of mutant individuals from the population, sx, equals the gain through mutation, u. But, if $sx = u$, then $x = u/s$. That is, there will be in the preantibiotic population of staphylococci a low frequency of mutant individuals containing an as-yet useless enzyme; the frequency of these individuals is determined by the rates at which they arise in and are lost from the population.

Numbers can be substituted for letters in the calculations of the preceding paragraph. The rate at which new mutant types arise is generally low; say one new mutant individual arises by chance among 1,000,000 normal ones ($u =$

1/1,000,000). Merely to have a number to plug into the calculations, I shall say that the reproductive ability of the mutant type is only 95 per cent that of normal individuals: $s = 0.05$ or $5/100$. Finally, assume that each of the 100,000 persons carries on his person 100,000 staphylococci; the staphylococcus population of the entire city equals 10,000,000,000. How many of these 10 billion bacteria are abnormal mutants? The equilibrium frequency of mutants is $u/s = 1/1,000,000 \times 100/5 = 1/50,000$. Furthermore, $10,000,000,000 \times 1/50,000 = 200,000$. Therefore, according to the several assumptions made here, there are a total of 200,000 mutants in the entire bacterial population or, on the average, two mutant bacteria per human being.

Time passes. Penicillin is discovered. To everyone's joy, staphylococci are susceptible to penicillin treatment, and it appears that severe infections and perhaps the organism itself can be completely irradicated.

Fate has been neither that kind to the human population nor that cruel to the bacterial one. The "useless" protein made by certain rare mutant bacteria is penicillinase, an enzyme capable of destroying whatever penicillin happens to penetrate the individual's cell wall. In the presence of penicillin the rare mutant is no longer inefficient in reproducing. Penicillin kills the normal individuals, thus leaving all of the staphylococcal living space to the mutant variety.

The introduction of penicillin therapy into the local hospitals and clinics and the sale of penicillin tablets on prescription at drugstores all over the city brings about a drastic change in the surroundings of, and the composition of, the local staphylococcal population.

Under the extreme conditions in which penicillin becomes a way of life for bacteria, the population gradually changes to one consisting almost entirely of the abnormal, penicillin-resistant type. The frequency of the previously "normal" type becomes extremely low; its frequency is now determined by the rate at which normal (susceptible) individuals arise by mutation and the degree to which their average reproductive ability is lowered by the now commonly used penicillin. More realistically, the final equilibrium proportions would be determined by the proportion of the total staphylococcus population that comes into contact with penicillin at any one moment. Not all persons in a large city are swallowing penicillin tablets or bathing in penicillin solutions continuously. Nevertheless, the formerly rare mutant type, resistant to penicillin, may approach an equilibrium frequency of 20 per cent, one resistant individual in every five, instead of the original frequency of one in 50 thousand. Indeed, among hospital employees – nurses, interns, orderlies, and ward helpers – the frequency of penicillin-resistant staphylococci may reach 80 per cent, four out of every five.

This is the nature of natural selection, the consequences of differential reproduction. The example has been couched in hypothetical terms but the development of resistance among organisms exposed to man-made chemical poisons has been an all-too-common experience. House flies and mosquitos, once easily controlled by a moderate exposure to DDT, are now largely unaffected by it. Instead, the massive treatments that are now necessary wreak havoc on organisms valuable to man. Body lice, carriers of the rickettsia that cause typhus, were controlled throughout World War II except under the terrifying conditions of the Eastern front. During the Korean War, however, body lice were largely immune to the effects of DDT. A malignant form of *Plasmodium falciparum*, the malarial parasite, which is resistant to the usual antimalarial compounds, is a serious threat in Vietnam and other parts of the world; a combination of chemicals has been devised that is effective in treating this severe form of malaria – at least for the moment.*

Higher organisms such as mice, men, and flies respond to the influence of an altered environment (to natural selection, if you will) more sluggishly than do microorganisms. Bacteriophages are small particles that reproduce in and destroy bacteria. If a test tube containing sterile culture medium is inoculated with a few bacteria one day, the tube will be cloudy and teeming with bacteria the next. If at that time a few bacteriophage particles are placed in the tube, it will be crystal clear on the following day because the bacteria have been destroyed (the phage particles are too small to make the medium turbid). More than likely, however, if the tube is kept for an additional day it will be cloudy once more. The few mutant bacteria that were resistant (immune) to the attack of bacteriophage will have replenished the bacterial population once more. Now, of course, the bacterial culture is immune to infection by bacteriophage.

The course of events does not go so simply in the case of higher organisms. In the first place, generation times are much greater for the more complex plants and animals. Furthermore, in contrast to bacteria, higher organisms generally carry genetic material obtained from both parents. Random mating between males and females of a population gives rise to many combinations of genes. In the formation of gametes, genes of maternal and paternal origins are shuffled to give an even more bewildering array of combinations. Individuals that survive to reproduce because of a particular genetic endowment may not give rise to offspring with the same endowment. The genetic basis for survival in such cases may be largely dissipated in the transmission of hereditary material from parent to offspring.

Progress under natural selection in the case of higher organisms does occur despite the slowing effect of gene recombination; fortunately, there is a

---

*To illustrate the speed with which optimism can change to pessimism, I have left the text of this essay (including the prophetic "at least for the moment") as it was originally written. I had used for reference the material that appeared in Science on December 20, 1968. Less than two months later, in the February 14, 1969 issue of the same journal, a letter appeared that reported that the malignant form of *Plasmodium falciparum* had spread from Singapore and Southeast Asia to the Philippines and northern South America. This letter also admitted that the malignant strains of Southeast Asia were no longer responding to the formerly highly effective combination treatment.

correlation in the attributes of parents and their offspring. Like, in other words, does beget like.

The well-known instances of change in natural populations within historical times have arisen most frequently in response to man's interference with his surroundings. The massive uses of chemicals in the control of insects has led to a large-scale destruction of certain species and to marked physiological alterations in the ones that have managed to survive (the target species in most instances).

The occasional adaptation that species of weeds have made to man's agricultural practices reveals the effect of natural selection in striking ways. A number of weeds have undergone marked changes in morphology and growth habits and, in this manner, have increased the probability that their own seeds will be included among the crop seeds harvested for next year's planting.

Regardless of the outcome of natural selection in any individual case, the point to remember from this essay is that natural selection is no more than a manifestation of a systematic differential reproduction of individuals that differ genetically. If the original population consists of types t_1, t_2, t_3, etc., and if these reproduce at different rates r_1, r_2, r_3, etc., then the composition of the population will change with successive generations. Under the influence of natural selection the population will evolve.

On the Coadaptation
of Hosts and Parasites

It is now appropriate to discuss a matter that has been hinted at in several previous essays, namely, the possibility that disease and man make mutual concessions to one another and that, if they coexist for an appreciable time, the nature of each is altered. They coadapt to each other's presence. The point is applicable to host-parasite and prey-predator pairs of all sorts as well as to man and his diseases. At the bottom of the matter is this: man and a disease-causing organism, a prey and a predator, and a host and its parasite are in each instance two component parts of a larger system, and in each instance the two components evolve in respect to each other within the larger system. Despite short-run appearances, the long-term existence of a predator, a parasite, or a pathogenic organism is utterly dependent upon the continued existence of a suitable prey, host, or other victim. Furthermore, since all species produce more offspring than their habitat can support, a selective attrition of the superfluous individuals is an absolute requirement. Predators contribute in a serious and worthwhile way to this attrition; their prey are dependent upon continued predation. The coadaptive changes in each component of the system are predictable in terms of differential reproduction and are expected to occur as a consequence of natural selection.

The following statement can be given as an axiom pertinent to contagious diseases: dead men do not spread disease. A virulent disease that multiplies rapidly within and causes the prompt demise of an infected person destroys any chance that it will be spread by coughs, sneezes, or personal contact. Even the time available for the infection or contamination of an insect vector is reduced. And so, if the individual organisms of a microbial population vary so that some of them cause rapid debilitation and prostration of their victims while others are less virulent and permit the patient to mingle with crowds of other, noninfected persons, the latter organisms will have the greater success in multiplying.

The greater spread of nonvirulent organisms in a population can be predicted by means of the arithmetic of epidemics developed in an earlier essay. Suppose a fatally stricken person infects 2 others before becoming bedridden; in contrast, suppose that a person infected with a milder strain of the same organism infects 4 others. During the second round of infection, the 2 fatally stricken persons will infect 4 additional ones but the 4 victims of the milder variety will infect 16. During the next round the numbers become 8 and 64, then 16 and 256, then 32 and 1024, 64 and 4096, and so forth. The less virulent form spreads much more rapidly than the lethal form.

The advantage of the less virulent microbial strain depends, however, upon the ability of infected persons to develop immunity against subsequent infection. In the absence of immunity, the figures cited in the preceding paragraph would merely describe the course of two illnesses that affect the population in tandem order. A recorder of medical history would write of such a population that a mild disease swept through the population and that shortly thereafter a slow-spreading but highly fatal disease attacked and killed everyone.

Immunity to the disease — perhaps we should say cross-immunity between the mild and virulent forms — guarantees that the slow-spreading, virulent form will not eliminate the population even at a deliberate pace. The fast-spreading organisms leave in their wake a collection of former victims who are now immune to the slower-spreading, virulent form. The virulent organisms disappear; their few early victims die before they encounter and infect the now-rare nonimmunized persons.

Selection within the human population proceeds at a slower pace than does that which takes place at the microbial level. A generation for man is 25 years; for bacteria it can be as low as 25 minutes. Suppose that men differ in their resistance to a disease, that they differ in a manner that affects the number of children they produce; if this supposition is true, then the more resistant become more numerous with passing generations. In today's Western civilizations where families tend to be uniformly small, differential reproduction does not appear to be a powerful agent for change. In societies where families contain enormous numbers of children and where disease and starvation are the chief regulators of population size, the greater reproduction of resistant adults following an exceptionally virulent epidemic need not be a trivial matter.

Specific examples clearly illustrating the hypothetical suppositions just outlined are difficult to find. The best illustration of these events is to be found in the attempted elimination of rabbits in Australia by means of the myxoma virus, the virus responsible for a highly fatal disease, myxomatosis.

The myxoma virus is responsible for an endemic disease of certain South American rabbits. It is a benign disease that is spread from rabbit to rabbit by mosquitoes.

In 1927 the virus was introduced accidentally into California and, presumably through transmission by mosquitoes, has been introduced into the native rabbit population of California. Both Californian and Brazilian rabbits of this account are relatives of the "cottontail"; they belong to the genus *Sylvilagus*.

Laboratory or hutch rabbits are not cottontails; they are "true" rabbits related to the European one, *Oryctolagus*. In these rabbits the virus produces a generalized disease that is nearly always fatal. This has been known since the late 1800's; in fact, the myxoma virus was isolated from stricken laboratory animals, not from wild carriers.

Following inoculation of a small number of virus particles into the skin of a European rabbit, an inoculation much like that caused by the bite of a carrier mosquito, the virus passes through the local lymph node into the blood stream. A general skin infection follows accompanied by swellings on the face and ears. Antibodies are produced by the infected animal although the virus kills so rapidly that these are difficult to identify in susceptible animals; surviving rabbits in the field carry antibodies and laboratory animals infected with an attenuated viral strain develop antibodies too.

Early attempts to control rabbits in Australia by means of the myxoma virus were not successful. At that time the role of the mosquito as an insect vector was unknown. In 1950, however, the disease escaped from the control sites and spread rapidly in the wild populations. There were further outbreaks in 1951, 1952, and 1953. Dead and dying rabbits literally covered the ground over wide areas of Australia; in these three or four years about 90 per cent of the entire rabbit population was wiped out.

In 1952 it was already obvious that the mortality of infected rabbits was not nearly 99.5 per cent as it had been at the outset of the epidemic. The mortality rate of infected animals has declined so that in recent years it appears to be about 25 per cent; three rabbits out of four now survive the disease whereas formerly this number was five or fewer per thousand.

Viruses isolated from wild rabbits have proven to be nonlethal when injected into previously uninfected wild rabbits. After two years of transmission in the field, less virulent strains could be isolated. These attenuated strains were also less virulent when tested on susceptible laboratory rabbits although at the time of the laboratory tests they were still fatal.

Australian rabbits, when injected with myxoma virus, develop symptoms about one day later than do laboratory animals. This may or may not be the result of selection for disease resistance in the local rabbit population; the records on this score are not complete.

Finally, the mere diminution of rabbit populations makes it less likely than in the 1950's that a mosquito will bite an infected and a noninfected rabbit in quick succession. The mosquito in this case is a passive carrier of myxoma virus; it serves as a flying pin. The virus particles from the infected rabbit merely contaminate the head and mouth parts of the mosquito and are injected quite by accident into the next animal that is bitten. According to the arithmetic of epidemics of an earlier essay, the sparse distribution of rabbits causes a decrease of i, the number of new infections that arise from a single old one. Anything that decreases the magnitude of i decreases the spread of the disease markedly because i is part of an exponential function.

All of the predictions that we would make on theoretical grounds seem to have been fulfilled in this massive man-made epidemic of the rabbit world. It is interesting to speculate, of course, that eventually the myxoma virus will be a benign disease organism of the Australian rabbit just as it is for the Brazilian cottontail. It is interesting, too, to note that this virus resembles closely the viruses that cause smallpox (a severe disease of man), cowpox (a mild one), and the common cold sore (herpes simplex). The herpes simplex virus of man, a virus that is frequently cited as the best of man's coadapted diseases, causes a fatal infection if inoculated into monkeys.

Rapid evolutionary changes resembling those undergone by epidemic diseases can be found occasionally in the case of animal or plant invasions. Animals or plants when introduced into a new geographic area, an area free of natural enemies, frequently spread with explosive speed and overrun enormous areas. The Japanese beetle, the Colorado potato beetle, the fire ant, certain marsh grasses, starlings, and the English sparrow — all of these have been overwhelming pests at one place or another in the past; some still are. The eventual outcome of these invasions is generally predictable: a predator or other "enemy" appears, the enormous population dwindles, and finally a reasonably stable equilibrium is established at which the predator and prey or host and parasite coexist. In the process, each of the two undergoes genetic changes that tend to lead to stable populations of both.

There is an exception to the predicted collapse of widespread invasions; this is related to the enormous acreages of farm crops maintained by man. The economics of farming demand the use of large areas for single crops. The Midwest consists of miles and miles of corn or of wheat. Orchards and vineyards extend as far as the eye can see in the Great Lakes region. Southern California and Florida have thousands of acres of citrus groves. In every respect areas that are devoted to single crops resemble those overrun at the height of an exotic invasion. In natural or wild populations the invading plant would be expected to dwindle under the impact of newly acquired, "natural" enemies. That this does not happen in the case of man's food crops is the result of great effort on the part of farmers and of the agricultural industry. The system is biologically unstable; it is kept going (like a plane in flight) only by the expenditure of energy. Left to itself — as we have seen in the case of Ireland's potato famine — modern agriculture would collapse. This point can be repeated in even stronger terms. As matters now stand, agriculture is in constant danger of collapse; in responding to the greater demands of growing populations, it becomes still more unstable. It is like building a house of cards — the probability of an eventual collapse becomes greater with the addition of each new card.

Some Letters Home

The following paragraphs have been taken from a number of letters written more than a century ago by relatives who had moved to Ohio from their family home in upstate New York. The correspondence has been included to illustrate the extent to which infectious diseases intruded upon human affairs during the last century, the very little that could be done to help the afflicted, and the individual resignation with which sickness and death were received.

Within the United States a steady reduction in the number of deaths attributable to infectious diseases began about 1900. Many believe that the miracle drugs – sulfa compounds and antibiotics – were responsible. These persons are wrong. Even before the discovery of these drugs in the 1930's and 1940's, an understanding of the natural history of microorganisms including those responsible for human diseases had drastically reduced the number of deaths caused by disease germs. On the other hand, wherever large numbers of persons congregate with inadequate sanitary facilities – army bases, fox holes and trenches, and refugee camps, for example – devastating epidemics are common even yet. The miracle drugs, which by now include many different antibiotics, have contributed to the world's population growth by permitting otherwise fatally stricken persons to recover and to reproduce.

Tallmadge, May 7th, 1852

Dear Sister,

I received yours of April 9th about two weeks since and was sorry to hear the sad tidings it contained. We see that grim messenger Death has no respect for age or rank; all are sooner or later to go to the house appointed for all living, it is therefore our duty to be ever ready at what time soever the summons is sent, whether in the morning, miday, or evening of our lives.

There has been a death in this neighborhood this week, an elderly woman, she left a large family but most of them are grown up, she had been sick about a week with the lung fever but appeared to be getting better and was able to sit up two or three hours at a time the day she died.

We are all enjoying tolerably good health at present. John has been afflicted with very sore eyes for more than a week but they appear to be some better; William Thomas was taken with them first but his are nearly well again.

Our family is reduced to the usual number four. Eliza left last week for Albian to attend boarding school six months or more. Jonathan boards with us but works for himself, he works the farm that was his father's and I believe he intends to buy it when the time comes that it can be sold, and I should like if he would go east and get him a good wife for I think he is worthy of one, and such articals as girls to make good housekeepers are very scarce in this country

• • •

You want to know what I am up to these days. Well I am not very much drove with business, at present the most I have to do is my house work and not a great amount of that as we have no milking to do yet so I get along without any trouble. Today I am baking half of a pig for dinner and should like if you had the other half or a whole one. You want to know what I think of Margret A's choice of a husband, I think she might done worse but if she is satisfied I have no objections to offer but think they will be rather an unhealthy couple to commence in life if they have to work for a living as the rest of us do.

•   •   •

Somonauk, Feb. 5th, 1857

Dear J.,

I set down to drop a few lines to you to let you know the state of our health. Two weeks ago I caught a bad cold which settled on my right breast, had to have a blister plaster which removed the inflamation, have some cough and raise some. Mary was take with a violent pain on her right side and breast with cold chills on Monday week. She is not much or no better has some feaver yet with headache.

Aunt J. came hear two days after I took my cold, don't know what we would have done if she had not been hear to waite on us and give the medicine.

•   •   •

I suppose you have heard of Margret's death. Uncle A.'s daughter, young A., was taken with some complaint but got better. The rest of your relations in good health. . . . We want you to write and give us the news and how Uncle Robert is, if in the land of the living, and how the rest of the folks are in Putnam.

•   •   •

Somonauk, June 18th, 1857

Dear J.,

Your of the third I received the eleventh and was sorrow to hear the loss of your son George. He was only lent to for a time to let you see that He that gives can call and take away in his appointed time to teach us to be submisif to His will and not put to much in our worldly informints. We have been enjoying C. good since I wrote you last. It has been general time of health in this place; there has been a number of deaths at the station east of this of children. Thomas B. lost his youngest child with the croop the 11 of May, five days less than one year old. Thomas has very poor health, had turn of the bilious disease, was confined home three or four weeks. Aunt M. is at George's. She has her health good for her. She calculated to come home with us on Sabbath night, we were out in the buggy and could not carry her. We have had a very wet week so far, the streams up over there banks.

Causes of Mortality in Man: Then and Now

Men have always died but for different reasons in different eras. Until very recently infectious diseases were the main cause of death. An understanding of disease germs and their transmission from man to man reduced the incidence of human sickness; the role of disease as a cause of death has been even further reduced since the discovery of antibiotics.

Infectious diseases struck their victims at all ages. In 1852 it was wise advice indeed to be prepared for departing this world in the morning, midday, or evening of one's life. With the virtual elimination of these diseases in the United States (and other countries of the Western world), death has been forced to seek other causes. Degenerative organic diseases are now the prime killers of man; these diseases, except for heart attacks, are as a rule diseases of advanced age.

The shift of mortality from infectious to degenerative disease has caused a drastic shift in the age distribution of the population of the United States during the past 60 or 70 years. The shift continues. The paragraphs that follow will, in the uninspired way of statistical presentations, cite numbers to support what has already been stated as fact. In addition, the two age distributions – those of 1900 and 1970 – will be examined further in respect to their bearing on the structure and operation of society.

One death in three at the beginning of this century was caused by one of the four communicable diseases: pneumonia, flu, enteritis, or tuberculosis. Strokes and cancers at that time were responsible for only one death in ten. Today, nearly one death in three is the result of cancer or stroke, and more than one death in three results from heart failure. Infectious diseases, other than some of early infancy, have disappeared from the list of the ten leading causes of death (see the accompanying tables).

The removal of infectious disease as a leading cause of death has had two consequences. First, population size has increased because of the survival to a reproductive age of nearly all children born. In the United States this particular consequence has been somewhat offset by a tendency on the part of many parents to limit their family size by artificial means. Where no such tendency exists, the assured survival of infants and young children has resulted in a veritable explosion in population numbers.

The second consequence of changing patterns of mortality is an unnatural distribution of age groups; this has been illustrated in the accompanying figure.

Within many if not most human societies, the age distribution diagram resembles a Christmas tree as does the portion of the figure representing conditions within the United States in 1900. The population of the nation at that time was essentially a population of young people; only one person in ten was over 55 years of age. Today, because death so often results only from degenerative diseases of old age, one person in five is 55 years old or older. From the ages of 15 to 75 the proportions of persons in succeeding ten-year age groups are very much alike. (The bulge caused by the excessive size of the 15-24 year age group in the 1970 distribution is not to be confused with the "Christmas tree" shape of the 1900 age distribution. In 1960 all age groups from and above 15 years were virtually identical. The bulge that is so apparent in the 1970 diagram will move upward into the higher age brackets and, should population control become effective, will in time cause a temporarily top-heavy distribution pattern.)

The shift in the age distribution of the American population has wide repercussions. Take, for example, the burden that the "able-bodied" segment of the population must bear. For sake of argument, let us say that "able-bodied" means ages 25-54 but let us omit from the "burden" persons aged 5-24 and 55-64. The burden, in this discussion, consists of the very young and the very old, below age 5 and, at the other extreme, age 65 and over.

If we accept the above definitions of able-bodied and burden, then in 1900 we find that 28 per cent of the population was able-bodied and 16 per cent constituted a burden. In 1970, 25 per cent of the population was able-bodied and 19 per cent were a burden. The ratio shifted from 2-to-1 to about 1.3-to-1; the responsibility per able-bodied man in 1970 was half again as large as it was in 1900. The high cost of medicare and social security arise in part from the high ratio of those who are to be cared for to those who are able to care and who (rightly) demand a living wage for their efforts.

The abnormal age distribution now characteristic of the population of the United States is responsible in many ways for the "generation" gap. In truth, we have a two-generation gap. In 1900 when the school children were told by their elders that "your school was good enough for us," oldsters of 40 years of age were talking to teen-agers; these oldsters had themselves only recently passed through the neighborhood schools. The number of 60-year-olds alive today is nearly as great as that of 40-year-olds. Teen-agers must now listen to the praises of bygone eras and of schoolhouses and of teachers of bygone eras sung in truly faltering voices. Listening may be painful but it is harmless. The failure of school districts to obtain approval for proposed budgets, however, is not harmless — at least in the long run.

The conflicts of interest between two equally numerous generations that are separated by still a third underlies a great deal of present-day unrest. The simplistic views of many of our older citizens, views that were obtained in a less complicated world, cannot adequately grasp the complex problems faced by today's youth. Many of our very senior citizens cannot understand the rate at which the world's face has altered, cannot see the enormous demands made upon man by his own sprawling cities and suburbs, and cannot comprehend the changes in the labor market made by automation and the use of new machines and electronic controls. In an age when it becomes ever more important, for example, for the young to learn fruitful and non-destructive uses of leisure time, outmoded traditions too often insist on teaching the virtues of hard work, often through the use of antiquated shop equipment that will never be encountered by the student again.

To point out the consequences of an abnormal age distribution in a population is much easier than to cite solutions for the problems that such a distribution causes. As in the case of many of man's problems, however, solutions will not be arrived at until the underlying biological facts have been recognized. Furthermore, the realization that society's problems, if left unattended, will get worse must strike home. This message rather than any numerical point that may have been made in the preceding paragraphs is the heart of this essay.

The Ten Leading Causes
of Death in the United States: 1900 and 1967

		Deaths per 100,000 Persons	% of All Deaths
1900			
1.	Pneumonia and influenza	202	11.8
2.	Tuberculosis	194	11.3
3.	Diarrhea and enteritis	143	8.3
4.	Diseases of the heart	137	8.0
5.	Cerebral hemorrhage	107	6.2
6.	Nephritis	89	5.2
7.	Accidents	72	4.2
8.	Cancer	64	3.7
9.	Diphtheria	40	2.3
10.	Meningitis	34	2.0
1967			
1.	Diseases of the heart	365	39.0
2.	Cancer	157	16.8
3.	Cerebral hemorrhage	102	10.9
4.	Accidents	57	6.1
5.	Pneumonia and influenza	29	3.1
6.	Certain diseases of early infancy	24	2.6
7.	General arteriosclerosis	19	2.0
8.	Diabetes mellitus	18	1.9
9.	Other diseases of the circulatory system	15	1.6
10.	Emphysema and related diseases	15	1.6

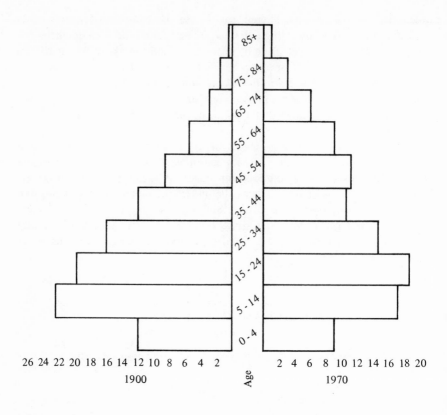

26 24 22 20 18 16 14 12 10 8 6 4 2   2 4 6 8 10 12 14 16 18 20
                    1900   Age   1970

The Biology of Cancer

Cancer kills about one American in six; that is, some 17 per cent of all deaths are attributed to cancers of various sorts. Only heart diseases outrank cancer as a cause of death. A half-century ago neither cancer nor heart disease could compete with infectious diseases for the grim honor of being the first entry on the mortality table, but now antibiotics and — to an even greater extent perhaps — knowledge about the biology of infectious diseases have reversed these standings.

In writing on the biology of cancer we are tempted to emphasize the dread with which it is generally regarded. Death by cancer is not a pleasant death. Only recently at a local town meeting I heard an understandably impassioned plea for a greater federal funding in the fight against this disease — the *leading* cause of death for mankind according to the speaker. She should, indeed, support cancer research but not, in my estimation, because it is the leading cause of death. There will always be a leading cause of death, and to believe that we can eliminate each one as it arises and takes its place at the head of the mortality figures is to pursue a will-o'-the-wisp. Indeed, in continually delaying the arrival of death we may easily find that the suffering of aged, dying persons is increased rather than the reverse. If the existence of cancer has been largely uncovered through the elimination of bacterial diseases, what next can we expect to uncover by now defeating cancer? Perhaps we *should* continue racing on the treadmill on which we find ourselves, but we should *not* continue racing without recognizing that we have in fact embarked upon an endless course.

I do not aim to dwell on the horrors of cancer. To whatever extent the horror of cancer can be discerned as something separate from the apprehension of dying, Tolstoy's description of the death of Ivan Ilych has been more than adequate. In this essay I intend to describe the obvious aspects of cancerous cells and discuss some possible causal agents underlying these external cellular characteristics. I must hasten to say, however, that the cause (if indeed there is but one cause) of cancer is still unknown, and so the essay must end in seeming disarray with many loose ends lying about.

Cancers are cellular growths; most persons are aware of this fact. The cancer surgeon attempts to remove these growths and their associated tissues. At times the surgeon performs an exploratory operation only to discover that the cancer is inoperable. The cancerous growths in this case are so widely spread in the body that it would be impossible to remove them all or, alternatively, some of the growths have seriously involved nearby vital organs with which the surgeon dare not tamper. To begin a discussion of cancer, therefore, we might consider growths and their appearance here and there in the body.

A cancerous growth arises from a single cell that, having undergone a hereditary change, multiplies by repeated division and gives rise to billions of descendant cells (billions of cells would result from as few as 30 or so divisions). The repeated multiplication of cancer cells reveals one of the ways in which they differ from most cells of the body. In the case of normal cells, multiplication is strictly regulated; cells divide only upon demand. Cells that line the intestinal tract divide and replace those that are continually sloughed off. Cells in the skin are continually dividing and replacing those that flake off as "dandruff." Cells in the hair follicles are also continually dividing. Tissues that are composed of dividing cells, like cancerous growths themselves, are sensitive to radiation, a phenomenon we encountered in the section on radiation biology in volume II. In tissues that are not continually undergoing a process of erosion and replacement, cell divisions are rare; they may occur, for example, only in the process of healing wounds. The uncontrolled and repeated division of cancer cells, therefore, is a truly extraordinary characteristic for a body cell.

A cancer cell also differs from a normal body cell in its indifference to the location of its site of multiplication within the body. Most cells of the body are components of tissues that are in turn components of organs and organ systems. The entire architecture of the body depends upon the strict adherence of each cell to its assigned role. Cancer cells are not only released from the inhibitions of self-duplication but they are also removed from the architectural restraints of other cells. Cancer cells can slough off from one growth, be carried in the bloodstream to other parts of the body, and upon coming to rest start dividing anew and producing new growths. Cells of individual growths can invade nearby organs, proliferate within them, and in so doing destroy the functions of the invaded organs. In an essay reproduced in the section on behavior in this volume, J. B. S. Haldane comments: "I shall die within a few years — perhaps one year if the cancer has sent a colony of cells to another part of my body, perhaps twenty-five if it has not. . . ." The cancer had spread; the abdominal surgery Haldane had undergone failed to remove all cancer cells, and Haldane died within the year he predicted as a possibility.

The differences between cancerous and normal cells can frequently be observed in cell cultures that are maintained artificially in laboratory dishes containing synthetic nutrient media. Even when growing in a glass dish on a laboratory table, normal cells multiply in number slowly and, in doing so, maintain an orderly relationship to one another. Colonies of such cells develop streamers and whorls that reflect the alignment of normal cells one against and parallel to the other. If, on the other hand, certain of these cells are treated in a way that makes them cancerous, they behave quite differently. They grow much more rapidly than the tissue from which they were obtained. They form irregular masses; under the microscope they are seen to move over and under each other with no tendency to set up definite spatial relationships with one another. In addition, we can make a test to reveal that these fast-growing, poorly organized cells are indeed cancerous. When injected into a test animal of an appropriate species (mouse and hamster cells are frequently used in tissue culture studies), the rapidly growing cells proliferate and form tumors; ordinary tissue culture cells when injected into a test animal do not produce tumors.

How is a cancer cell produced from a normal cell? This is a difficult question and one over which there has been considerable controversy. There is, in fact, some doubt as to whether all cancer cells in the body of a cancer patient need be descendants of a single primary cancer cell. In the tissue culture cells just described, the experimental circumstances are quite clear. The cancer cell arises from a normal cell following infection by a virus; in this case, one cancer cell arises from one normal cell, and the causative agent is the virus particle. These facts are established by experimental techniques that have proven invaluable to the theoretical studies of molecular genetics. An entire battery of laboratory procedures now exist for studying these experimental cancers in exquisite detail.

Whether all cancers are caused by viruses is not at all certain. A number of them clearly are. A recent flurry occurred over the authorized sale of all but the most repugnant chickens among those suffering from leukosis, a cancer of poultry known to be caused by a virus. The ubiquitous herpes virus of humans (the cause of the cold sore) is suspected of causing cancer of the cervix in women; uncircumcised men are the suspected carriers in this case. A number of cancers in mice are known to be transmitted from mother to offspring (or from female to foster offspring) by viruses that are found within the mother's milk. A simian virus that was inadvertantly permitted to contaminate early polio vaccines is known to cause cancers if it is injected into young hamsters; no evidence has appeared, however, to suggest that cancers attributed to this virus have appeared in inoculated children.

A lengthy (and at times lively) discussion has occurred over the past half-century concerning the possibility that cancer is the result of somatic mutation, that is, of a genetic change within a body cell. Geneticists have tended to favor this view. It has a certain appealing simplicity. It involves quantitative concepts, both in the probability of appearance and in the subsequent course of the disease, that lend themselves to mathematical manipulations. And it encourages rather precise experimentation where expectations can be accurately stated for comparison with observed facts. One of these expectations involves the increased incidence of leukemia among persons exposed to X rays or to atomic radiation. The somatic mutation hypothesis of geneticists gains support, in fact, from the frequently observed ability of mutagenic agents to produce cancers and of carcinogenic (cancer-producing) agents to produce gene mutations.

The simplistic views of geneticists have not gone unchallenged by the medical men who specialize in cancer therapy and cancer research. Their objections are based largely on complications that exist but which need not be expected according to simple genetic models. These complications include, for example, the complex tissue pathology that precedes the origin of certain cancers. Treatment of the skin by carcinogenic chemicals often results in the formation of recognizably precancerous cells, which are clearly abnormal to a pathologist, from which, and only from which, a cancer will eventually arise. The origin of a cancer cell apparently need not be a single-step affair.

Many of the points of contention of past discussions over the causes of cancer may be resolved by modern research on virus-induced cancers that are produced experimentally in tissue cultures. In some instances a simple model encounters trivial complications; for example, a million (not just one) artificially induced cancer cells must be injected into an experimental animal in order to produce a tumor. Once this fact is understood, it does not cause undue hardship for the experimentalist. Certain intestinal infections, to illustrate by analogy, arise only following the ingestion of a million or more causative bacteria, but this fact has no profound modifying effect on the germ theory of disease.

In many respects the events that are associated with viral infections mimic those that are based on genetic events. The reason for the resemblance is that the infective portion of a virus particle *is* genetic material, and its functioning within a host cell calls for its operation within the gene-controlled machinery of that cell. At times the infective particle merely reproduces large numbers of viruses that "leak" out of the cell slowly and infect nearby cells (the encephalitis virus behaves in this way) or "burst" the cell upon escaping (as bacteriophage rupture bacterial cells at the end of the reproductive cycle). At times, too, the infectious particle enters into the genetic apparatus of the host cell and from that moment reproduces with the cell's genetic material at each cell division. Virus particles that have been incorporated in this way are normally undetected, but they can often be released from their sites of incorporation by X rays and other types of mutagenic radiation. Were these formerly undetected viruses upon release to be responsible for cancer-cell formation, the incidence of induced cancers would increase hand in hand with increased exposure to radiation exactly as gene mutations themselves.

It is not for me in this essay to resolve the dispute over the cause of cancer. Instead, it is my task to describe the biology of cancer. A cancer represents the transformation of a portion of one's body from an organized to an unorganized state. The disorganized portion, the cancer, steals nutrients from other cells. By stealing from other cells it lowers the resistance of the body to bacterial infections. Upon invasion and destruction of a vital organ, it causes the death of the very body of which it is a part. Perhaps nothing illustrates the distinction between mind and body more clearly than the reference of a cancer patient to his cancer as "that thing." And perhaps nothing better emphasizes the delicate balance that exists between the various cells and cell systems of a properly functioning body than does the havoc wreaked by those very same cells once they have escaped from effective control.

Death as a
Biological Phenomenon

Of all animals, human beings are seemingly unique in the realization that death is the terminus of each individual's life. No other animal, as far as we know, has the foresight to anticipate its own demise. Other animals may recognize momentary danger; they may recognize the need to assist others that are in danger; but they do not recognize their own eventual death.

Every individual living thing is mortal. In the case of household pets and farm animals, direct observation reveals that these grow senile and die despite intensive care. Wild animals would also die of "natural" causes if accidents and death by predation did not intervene. There is some question as to whether our routine observations are pertinent in the case of certain trees. Does, in fact, a giant redwood have a finite life span? I believe so; I suspect that redwoods have an ever-increasing probability of dying so that a full-grown tree becomes less and less likely to survive each passing year.

A remarkably simple relationship exists between death and reproduction. An immortal organism could not afford to reproduce if its offspring in turn were immortal. A constant input of individuals through birth and no corresponding outgo through death would lead to a cluttered world indeed. On the other hand, a long-lived organism that retains its reproductive ability undiminished throughout life can afford to eliminate its own young upon birth. This is essentially the pattern followed by the giant redwood; the only surviving seedlings in a redwood forest are those that germinate at the site of a recently destroyed tree. Only by the ruthless elimination of its young could an immortal form retain its ability to reproduce but, in such a case, "immortal" would mean not immortal but long-lived.

Of the four combinations of "mortal-immortal" and "reproduce– nonreproduce," only organisms that are mortal and reproduce can exist and be recognized as living. Mortal forms that fail to reproduce would quickly disappear. Immortal forms that reproduced would have already cluttered the world intolerably. Immortal forms that did not possess the power of reproduction would be exceedingly hard to recognize; presumably, such a unique form would of necessity assume characteristics not generally associated with life as we know it.

The world has undergone many changes since life first arose. Oxygen has accumulated in its atmosphere; by the same token, many noxious (to life today)

93

gases have disappeared. Enormous glaciers have appeared and receded only to reappear. In the past, tropical climates have existed where today's climate is subarctic. Deserts have appeared only to disappear, at times under newly formed inland seas.

Throughout this unending change, life has continued to exist. The secret of life's success resides in the dual abilities: (1) self-reproduction and (2) change, including the ability to reproduce the changed form. These abilities reside in the genetic material. They have undoubtedly been a characteristic of life from the very beginning. The primitive molecules that were the first forms of life presumably reproduced themselves very inefficiently, but decomposition was also a slow process, dependent largely upon the thermal instability of complex molecules. The rate of decomposition would have speeded up as one molecule assumed the ability to utilize others that were themselves reproducing; in turn, however, this ability would correspondingly speed up the rate of replication of the "predator" molecules. In retrospect, we would say that death in the form of destruction by accident or by assimilation has been a feature of living systems from the onset of life on earth.

Although the specifics may be difficult to enumerate, the average length of life for a given species must in general be subject to modification by natural selection. The more complicated the organism, the longer the embryonic stage during which it is assembled. Obviously, the parent must survive long enough to reproduce. In higher animals the young frequently continue to depend upon and learn from their parents long after birth. Again, in this case there would be strong selection favoring a life span sufficiently long for parents to complete this period of adolescent training of their offspring. In the case of animals living in groups held together by social bonds, oldsters must survive to complete their contributions to the upbringing of the younger generation. There is, however, very little selection among animals for the prolongation of the life of parental individuals beyond the reproductive years and the educational period of the young. On the contrary, a case could be made that populations of animals within which old individuals disappear when their usefulness terminates will be more successful than other populations that are hindered by the burden of the aged. A great deal of evolutionary change comes from interpopulation differences such as this. Natural selection does not operate solely on an individual *versus* individual basis within local populations.

Death, therefore, has its place in the biological world. Without it, the ability to reproduce would be an absurdly superfluous ability. The ability to change, and hence to evolve, would be impossible in the absence of death. Life, as we know it, exists only because the existence of individual organisms is finite, because death eliminates what *has been* successful yesterday to make room for what *may be* even more successful tomorrow.

On the Dignity of Dying

The inclusion of this topic in a biology text is prompted by a number of considerations. Medical science has advanced to a point where, at least in individual cases, dying has become extremely difficult; the machinery that can be called upon today to maintain a semblance of life in a failing body is tremendous. Each of us must ponder the consequences of medicine's technological advances. The problem confronting us has been stated bluntly in these words: "When do we pull the plug?"

A second consideration is that illustrated by the passing of Ivan Ilych. A man, happily married with children and good friends, is stricken by an inoperable cancer. Other than the simpleminded servant boy, Gerasim, none of the characters in Tolstoy's novel could bring himself to admit in Ivan's presence that this man was dying. Weeks passed during which plans for the children and wife could have been made with Ivan's active help, plans that would have set his mind to a useful and satisfying task but, as in many real instances, these weeks were allowed to pass in a bitter, badly acted charade.

A last consideration, the one that prompted the title of my essay, stems from the obverse consideration, the respect for life. Instilling in youngsters a respect for life is one of the reasons for teaching biology in elementary grades. This respect need not take the extreme form of Dr. Schweitzer's reluctance to kill any living thing. It can be merely an acknowledgment that a living organism, even the simplest and most lowly, is a marvelous machine and that it is not to be destroyed senselessly. From Scottish ancestors the admonition is passed down as "Don't break what you can't make." If life in truth deserves respect, then it follows that life's termination deserves dignity.

Following the decision to include an essay on the dignity of dying in this collection, a colleague suggested that Hemingway's *Death in the Afternoon* might be a useful source for material on the biology of bullfighting. In this book Hemingway has written an essay entitled "A Natural History of the Dead." For the most part it deals with the biology of decomposition. Now, decomposition does not interest me. After life ceases, the fate of the remaining physical matter is of no consequence other than for the role it plays in the food web of life. The deference normally accorded a lifeless body is not accorded it in order to maintain the dignity of dying but to show respect for the memory of the deceased. In Hemingway's essay, only the concluding exchange between the doctor and the lieutenant touches on the dignity of dying as I use these words.

At various other places within *Death in the Afternoon* Hemingway touches again on matters that are related to what I term the dignity of dying. From his comments on the mortally wounded bullfighter Gitanillo it is obvious that Hemingway regarded prolonged dying as shameful or degrading. A prolonged death in his eyes lacked dignity. It is not entirely surprising then that Hemingway ended his own life as the indignity that accompanies (in his view) the infirmities and inaction of advanced age approached.

Death is a biological necessity. All living things are mortal. That each will die is certain; only the cause of death is uncertain. During man's history the main causes of death have changed from infectious diseases to organic failure. Persons rarely catch fatal diseases purposely; no one deliberately suffers a debilitating heart attack. And so where is the shame or the indignity? In this respect, Tolstoy's Gerasim cheerfully caring for the helpless Ivan Ilych radiates a calm strength quite different from the fearless daring we associate with Hemingway, who nevertheless was seemingly repulsed by dirty sheets.

At the outset we asked, "Who will pull the plug?" To what lengths should society go to maintain respiration, circulation, and kidney function mechanically while feeding the terminal patient intravenously? When should these costly procedures stop? Who should decide?

I have commented on the inability of Ivan Ilych's family and friends to talk to him of death. "What tormented Ivan Ilych most was the deception, the lie, which for some reason they all accepted, that he was not dying but was simply ill, and that he only need keep quiet and undergo a treatment and then something very good would result. . . . This deception tortured him – their not wishing to admit what they all knew and what he knew, but wanting to lie to him concerning his terrible condition, and wishing and forcing him to participate in that lie. Those lies – lies enacted over him on the eve of his death and destined to degrade this awful, solemn act to the level of their visitings, their curtains, their sturgeon for dinner – were a terrible agony for Ivan Ilych." They were lies that made a mockery of his entire life.

What is a person? Who was Ivan Ilych? Who are you? Who is the person sitting next to you? Not until these questions are answered, do matters of the sort discussed in the two preceding paragraphs come into focus. These questions may not be answered satisfactorily in this essay because they are not simple ones. On the other hand, they will *never* be answered unless we strive to see them clearly.

Who am I? Am I merely two arms, two legs, a torso, and a head? Obviously not. Legs and arms can be amputated and discarded but I remain. Large portions of my internal organs can be removed and still I remain. My heart can be exchanged for another one or for a mere mechanical device and I am here yet. The only place where I exist as an entirely unique person different from all others is in my brain, in my mind. All else, subject to the surgeon's skill, can be removed and replaced by plastics, by motors, or by transplants from organ banks. Transplant my head, however, and I go with my head. That must be the inviolable rule when such transplantations become possible: the person goes with the head.

The mind, despite the objections of many, is a computer. It is not "just a computer," if we may paraphrase the common objection to evolution as "just a theory"; it is an exceedingly marvelous computer. It is a computer that can be programmed by way of any one of five senses individually or by any combination of them. Programs devised and stored in one computer (the teacher) can be transmitted to another (the student) by the spoken or written word or by diagram. The insertion of new programs and the revamping of old ones in the mind goes on year after year. All of one's experiences — school, work, play, home, in town, abroad — are registered and worked into a massive memory bank where they are available in most instances for instant recall. Moreover, it would appear that the mind is constantly reprogramming its memory bank, not only by making the existing programs more efficient but also by reordering them in order to avoid accidental impairment through the irreversible loss of nerve cells in the brain. There are roughly 10 billion neurons in a human brain at birth; between the ages of 30 and 75 this number decreases by nearly one-third, a loss of some 200,000 neurons per day.

The unique aspect of the mind as the "home" of the person cannot be overemphasized. There are some who, in anticipation of sophisticated experimental techniques, would duplicate favorite children by the transplantation of nuclei from the child's body cells to enucleated egg cells. The new embryo would be genetically identical to the donor child. At least one father has said that he would retain a living culture of his son's cells if the boy were to enter the army; in this way he would be able to "duplicate" him in case he became a war casualty.

These dreams are nonsensical. The grown son is an individual in the sense and to the extent that his accumulated experiences differ from those of all other persons. No parent of a twenty-year-old son will start with a new baby — even if it is genetically identical to the first — and lead it through the same experiences and to the same end that had been reached before. The physical similarity between the new and old versions may be remarkable but the two individuals will be different persons just as any other two persons are different.

Aside from occasional passing interest in and pleasure from his physical body, the bulk of all that is interesting about a person is in his mind. Here is the center of all that is genuinely worthwhile about him. And here in large measure lies the solution to the questions raised by the artificial maintenance of life. Life is obviously maintained as long as the mind is alert. When tests of the mind reveal that it is no longer functioning and that the clinical apparatus is merely maintaining a collection of body cells, the apparatus should be disconnected and readied for better use. These, however, are the obvious points to which nearly all persons would agree. More difficult questions arise (in fact, they are here already) when too many patients require the use of the same apparatus. Is "First come, first served" always the best rule? How does one establish priority on the basis of age, occupation, sex, mental condition, family status, and the like? Questions of this sort cannot be answered as easily as they can be raised! We can be sure that they will not disappear; on the contrary, they will be clamoring for more and more attention from now on.

Questions concerning the termination of life are difficult, in part, because of the extreme care with which they must be both phrased and answered. As a personal rule of thumb, I maintain that one person (or society as a whole) should not cause another either physical pain or mental anguish. Casual discussions of "pulling plugs" can easily cause the seriously ill and aged persons extreme mental anguish. It would be unfortunate, indeed, if the hospital were to be looked upon as a place where life may terminate not only because some illnesses fail to respond to treatment but also because medical officials might withhold adequate treatment. To alleviate fears that may arise through misinterpretation or misunderstanding, I must repeat that in my view life has ceased when the mind ceases to function, that is, when appropriately sensitive tests fail to reveal brain waves or any other sign of mental activity over an extended period such as 24 or 48 hours. Many persons enter the hospital knowing full well that they will not leave it alive. It would be shameful in the extreme if these persons feared that their stay was to be deliberately shortened by those persons whose task it is to care for the ill and the helpless.

The claim that the mind is the person (or at least the truly important part of the person) tells us something about Ivan Ilych's associates. Their actions — and how common such actions are — made a mockery of all that was important about him. Physically, he was failing; he suffered pain, his organ systems were malfunctioning, and he had insufficient strength to care for himself. Nevertheless, until the very end itself, his mind was normal. It would have functioned with a satisfactory purpose had anyone helped or encouraged it. Instead, it was presented exclusively with misinformation — nothing is wrong, you will soon be better, stay quiet, take your medicine. Ivan was eased out of existence by chatter about oysters and cards. Elementary-school children are taught a respect for life. Tolstoy has described the ultimate in disrespect — confounding death with indignity.

Of the dignity of dying, as the obverse of a respect for life, there is little that we can say that is defensible in rigorous terms. We may make points that are instructive but their support must be argumentative rather than definitive. It seems, however, that if we do respect life — perhaps through awe for its complexity and intricacy — we should look upon its termination as more than a casual happening. This certainly is the spirit that the new biology texts are attempting to instill in children. Experiments with worms, insects, frogs, fish, and even mice and other mammals are carried out in order to learn, not to harm. Children cannot be expected to carry out pioneer experiments; on the contrary, they must repeat well-known ones of the past. Nevertheless, the knowledge they gain is valuable. In doing an experiment, if they are to show respect for the experimental organism (even an earthworm or a grasshopper), they must do their homework thoroughly in advance; this gives them the maximum information from the least killing. If children are taught otherwise, they are taught to be butchers, not experimentalists.

And so, too, with people. The dying deserve respect and decorum from those about them. They deserve a sense of purpose in having lived and in taking leave. A senseless killing is repulsive because it serves no purpose. We can argue that purpose was not lacking in ancient sacrificial ceremonies. We can even devise a justification for the prompt executions of olden times where an execution served as a warning for others. The promptness of the execution was essential to focus attention on the warning and to keep the execution from being an entertainment.

With the preceding notion, in mind, it appears to me that a judicial system that safeguards the rights of the accused and permits him the right of repeated appeal cannot possibly carry out executions with dignity, respect, or a sense of purpose. In the case of capital punishment within a democratic society, the very devices that provide the accused with time for appeal also provide time for the population at large to develop a carnival air. In a case of that sort the state cannot act with dignity, only with petty revenge. In a democratic society, death with dignity by capital punishment is rendered impossible by the defendant's right to appeal. Because I regard the right to appeal as sacred, I argue — and this argument terminates the essay — that capital punishment should be abolished.

Suicide
and Man's Right to Die

Beyond the onslaught of childhood diseases, beyond the accidents of adolescence, and beyond the bacterial and viral infections of young adulthood and middle age, mankind has come upon the degenerative diseases of old age: heart ailments, strokes, arthritis, cancer, and senility. Yet the search goes on: for cancer cures, for successful organ transplantation techniques, and for mind-restoring drugs. The search goes on as if individual men were destined to be immortal. It goes on as if the penultimate day of one's life is a transplanted happy and carefree day of one's youth.

At any moment much of the fight against disease is aimed at the leading cause of death. This is a fruitless, endless fight. As long as persons are mortal there must be deaths and, as long as there are bookkeepers and accountants interested in body counts, one type of fatality will be billed as a leading cause of death. The fight to prolong human life (or is it to delay dying?) will be never ending if mankind wishes to continue waging it.

The 20th century has been a century of economic reform. A living wage is now generally recognized as a proper reward for the labor of each working man. Sanitation men have starting wages and opportunities for promotion comparable to or better than those of most municipal employees. Nurses who not long ago ate in the kitchen with the household help when on private duty are regarded as professional workers today. Personal services of all sorts are expensive in a society that recognizes that each worker deserves to live in reasonable comfort. To the dismay of those who would promote tourism in the United States, most travelers from abroad cannot afford the rates of American hotels. The reason lies in the need for good hotels to provide many employees – bellhops, maids, waiters, waitresses, and bartenders – to care for the customer's different needs. And personal services in the United States are dear.

Old persons need a great deal of personal attention. Old persons, if they are wealthy, can afford adequate care. Those who are poor cannot. Patients residing in nursing homes, in effect, pool their financial resources in order to have attendants at their disposal. The cost of labor is so great, however, and the amount of money available to most aged persons so small that in many nursing homes the level of care is depressingly substandard.

A society that strives to prolong the life of each individual while arranging that personal care for the aged and infirm – for those who need it most –

becomes a financial impossibility should reexamine its values. That the ever-increasing prolongation of life would emerge unscathed from such a reexamination is by no means assured. A colleague has suggested (not entirely in jest) that the best investment for heart research funds would be a careful study of the past lives of those persons who have succumbed to sudden, massive heart attacks in their mid-sixties. The rest of us could then be advised how to live so that we, too, might die in the same way at roughly the same age.

The aim of this essay is to raise the issue of suicide as a socially recognized and sanctioned – a permissible – avenue of escape for those who are either terminally ill or merely bored with the tedium of advanced age. The issue involves the commonly held view that a would-be suicide is *by definition* insane and therefore unfit to decide his own fate. Rephrased, this question becomes, "Can suicide be man's last rational act?" The issue also involves the economics of intensive hospital care and of prolonged care as a semi-invalid. And finally it involves the notion of freedom, the freedom to do as one wishes with one's self. To deny this freedom in any degree risks making the individual a slave to a corresponding degree.

Persons who have led full, active, and productive lives often find that those things that in earlier years gave them pleasure are now suddenly beyond both their physical and mental ability. Some persons in anticipation of retirement years cultivate secondary, less demanding skills for use and enjoyment during the final period of their lives; these are the lucky ones who reap the joy of relaxing with their hobbies. Others are not so farsighted; they assume that their abilities will last forever or, more likely, that they will die in the harness before the onset of physical debilitation. Among the latter are the James Forrestals and the Ernest Hemingways of the world. For these persons a world in which they play no role palls, and they want to get out of it. Are the desires of these persons – persons greatly respected for much of their lives – now to be denied? Are strangers to urge them to reprogram their lives so that they raise flowers as an escape from boredom? If these once active persons argue that death is preferable to sustained boredom, must this attitude be interpreted as a sign of insanity? I cannot believe that it should be so.

The period immediately preceding life's termination is the most costly of a man's life; tuition and other expenses of college years are a pittance in comparison. The hospital, especially its intensive-care unit, within a week or a month can exhaust the modest savings of an elderly person or of a young couple. If we restrict ourselves to patients with terminal illnesses, is there any reason why, at a time deemed appropriate by the patients, suicide should be looked upon as an unreasonable or senseless act? By "terminal" illness I mean one that will proceed inexorably to death. There is, to be sure, a remote chance that a last-minute cure will be found that will arrest the advanced cancerous growths or will repair the grossly damaged heart muscle of the individual patient. There are some who argue against self-destruction on the grounds that within the hour such a cure might be found. Those who advance this argument seriously, in my opinion, reveal a more devastating lack of judgment that do those who contemplate suicide. The slightly premature extinction of one's life during the last stages of a terminal illness is for many persons an exchange of several empty days or weeks for the knowledge that their dependents will be financially secure; this knowledge vanishes if one submits his being to intensive care. For many, suicide is a release from the fear, as one British physician has said, of what doctors can and will do in an intensive-care unit (plus, in the United States, what they will charge for doing it) and of a vegetable life to be maintained in a geriatric ward.

When I say that a person under certain circumstances may be permitted to end his own life, I mean that he is entitled to professional aid and advice rather than be subjected to physical restraint. By granting this permission we raise, however, the matter of deciding who is to be permitted to commit suicide and who not. A child obviously is not to be permitted because children are supposed to be incapable of making mature judgments. Terminal patients approaching the final days of their lives are to be permitted. The elderly person who is bored to distraction in retirement and who has no desire to be reprogrammed for a new way of life is also to be permitted. The position I have taken is that it is *his* life to do with as *he* wishes and the mere desire to end it must not be interpreted as a sign of insanity.

Left, however, are the multitude of young adults — students, workers, and housewives — who commit suicide each year. What is their standing in relation to my stated position? Throughout this essay, I have discussed suicide as a rational act or, to be more precise, as if it can be a rational act. Acts of self-destruction committed impulsively and irrationally do not fall within the scope of this essay. If, however, a young person were to approach me and propose suicide as a solution for his problems and if he were to advance rational reasons in support of his decision, I could only attempt to dissuade him with rational counter-arguments. Were I to fail and were he to persist in his views, I would have to concede defeat. To do otherwise — to pounce on him and restrain him physically, for example — would be to make him my slave, to steal some of his freedom. The right to commit suicide, therefore, like the right to marry (in most states) and the right to an abortion,* would be attached to a demonstration that the decision has been based on sustained — and, presumably, rational — thought. The "demonstration" in the case of marriage is frequently given in the form of a 72-hour waiting period between the declaration of intent and the act itself. For both abortion and suicide, official sanction might require that the person undergo an intensive but sympathetic interview with a counselor at some time during the enforced waiting period. At the end of the period, however, the girl (in the case of a desired abortion) or the mature person (in the case of the would-be suicide) must make and bear responsibility for the final decision.

*See the essay entitled "On Population Control" in Vol. I.

Society's responsibility in respect to acts against one's self must be restricted to insuring that such acts, if they are to be condoned or in some instances aided, are rational acts. Society can do no more; neither can it do less.

SECTION TWO

Sex

Introduction

In the matter of race, we took on an explosive issue; in sex, we take up a formidable one. What can I write for a generation of students whose biology texts since the seventh grade have dealt frankly with both the anatomy and physiology of sex? Am I to deal with techniques? Surely not! To paraphrase Dr. Albert Szent-Györgyi,* at my age I do not feel I have the qualifications to talk on sex today or pose as an authority on the subject. If the reader feels uninformed about matters of technique, let him try either the movies or the books advertised in the Sunday *New York Times*. What can I write for the predatory coed who, in an effort to win the more serious attention of a casual aquaintance, boasts that at least a dozen fellows are qualified and willing to attest to her skill in bed? On second thought, for that girl I may be of some help because information on the current venereal disease epidemic might be highly appropriate.

And how do I choose a passage from a book in an effort to bring observations on sex into the classroom? Do we look through the eyes of a Portnoy? Possibly, but I have sought other eyes. Even though it deals with another century, I have cast my lot with "Theme in Shadow and Gold," a chapter from Carl Sandburg's *Always the Young Strangers*. In some eighteen vignettes Sandburg reports with sympathy and candor on the love lives and sexual mores of Galesburg, Illinois, during the late 1800's. I think there are some worthwhile messages here and there but, even should I be wrong, the selection makes for an evening of relaxed reading. So, enjoy it.

The appealing features of Sandburg are his gentleness and his honesty. He is frankly baffled by a childhood friend who regarded sex as loathsome. Sandburg has simple tastes; he loves simple, pleasant women and healthy babies. Yes, babies are especially nice – like a syrupy medicine.

Much of Sandburg's Galesburg is reminiscent of small towns as late as the 1930's. Fouling the boys' outdoor privy at the Middleboro (Pa.) Grade School was an important topic for teacher-student discussion; sex was not. Still the sex of the 1930's – like that of the 1800's, of the 1970's and of all time – could bring babies or spread disease, could cause pleasure or sorrow, could be lustful or adulterous, as well as tender, and could occur between growing children as well as between mature, work-worn couples. Much of life has become more informal and less restrained, but sex has not changed a great deal – despite what some, both young and old, think.

*Albert Szent-Györgyi, *The Crazy Ape* (New York: Philosophical Library, 1970).

"I have," Sandburg confesses, "an album of faces in my memory." So do we all, although at college age we do not readily admit it. This album of faces is one of the features that makes each person unique; its existence is one of the bases for my repetitious argument that the person is the mind whereas all else is but tissue, much of which is either removable or replaceable by a sufficiently skilled surgeon. In the mind there are always thoughts to be thought and work to be done, but in rare moments of relaxation we can turn to this marvelous book whose pages can be flipped at will, this album of faces of the very special people of our lives.

Theme
in Shadow and Gold*

Carl Sandburg

Earnest Elmo Calkins so loved Galesburg and Knox that his book They Broke the Prairie can stand as a classic portrait of a small town and college in the Midwest from 1837 to 1937. In a section on "Culture" he mentions physiology being taught in both school and college" as if the two sexes were identical and lacked reproductive functions." He quotes a disgusted senior: "You can go clear through Knox College and never learn that babies are not found under gooseberry bushes." He gives "an embarrassing incident" from high-school days:

The Shakespeare used was Hudson's edition, expurgated to protect the morals of high-school students from the facts of life. The worldly old Polonius was commissioning Hamlet to inquire discreetly into the possibly gay life of his son, Laertes, in Paris,

"I saw him enter such and such a house of sale."

What was a "house of sale"? The next line had been excised by Hudson, and no one could explain the phrase until a student who had an unexpurgated edition read the next line — "evidently a brothel." There was a painful silence; some of the boys snickered; others evidently did not know what the word meant. The teacher recovered first, and with heightened color said, "We will now go on."

As to the pure and the impure observed by Calkins in the 1880's and 1890's, he wrote: "An expectant mother was a disgraceful spectacle, followed by whispered comment. Childbirth was never discussed, either in school or society. Children imbibed no beautiful associations with maternity. The obscene passages in the Bible were always skipped in the Sunday-school classes." Once a year the high-school boys were called into a session for males only — but not for a talk on sex hygiene. "No, the lecture was a stern admonition against fouling

*Abridged from Always the Young Strangers, copyright, 1952, 1953 by Carl Sandburg by permission of Harcourt Brace Jovanovich, Inc.

the public privy and writing on the fences with chalk words now found in current novels. The Puritans who warred against slavery, intemperance, Sabbath breaking, with beating of tom-toms, evaded the subject of sex completely. Whatever a child learned was from contemporaries, and it was years before most of them got the matter straight in their minds."

•   •   •

A gentle human soul was Edward Higgins. His father and mother were out of old-line Pennsylvania Dutch and New England stock. The father had a good job at the Brown Cornplanter Works. They were Methodists, pious but not sanctimonious. They said grace at the table. They gave their five children what is termed "a good up-bringing," sending them to high school and keeping them fairly regular at church and Sunday-school attendance. They had books in the home, and besides the Galesburg Republican Register, they took the Chicago Tribune, the Youth's Companion, St. Nicholas. The father voted the national Republican ticket and was a total abstainer, voting in the local elections the No-License ticket which aimed to dry up the saloons. They were a clean American family, the atmosphere healthy and wholesome. I delivered milk to them, came to know them, and liked them. There were cold winter days when Mrs. Higgins asked me into the kitchen for coffee and fresh doughnuts.

Edward was the oldest of the five children. He finished high school, became a certified public accountant, and half the time was out of town on business.

Then a blow fell, one of those things beyond prediction before it happens, almost beyond belief after it is a fact staring you in the face and saying, "What's to be done about it?" In the realm of relations between man and woman is a platitude that anything can happen and usually does. Edward Higgins, a bright and good-looking fellow, with a reputation for being "honest and upright" and keeping company with decent and respectable people — Edward Higgins had suddenly married a woman. And they had set up housekeeping in a large house on a respectable North Side street, and the woman Edward married was known as one of the sorriest, blowsiest whores that had ever hit the town. There too to that house I went with my milk can and poured out the quarts for the loose-mouthed, slipshod, barelegged, loose-gowned slattern who ran the house and while Edward was away on business trips had men and drinking parties in the house. I saw this affair going on and Edward's mother sad-faced about it. I moved on into another job and didn't hear how it came out, though I am sure Edward couldn't have stood it for long. One night when I passed the Higgins house and saw their lighted windows I said to myself, "You can read happiness or anything else you want in the lighted windows of a house at night."

Once in a blue moon, possibly more often than the appearance of a white crow, a lyric love may arrive to a prostitute and a man's need of her. My good

Swedish friend, Dr. Thor Rothstein, with a reputation beyond Chicago in his day in the medical profession, told me of a case. The man had syphilis. The woman who had given it to him he wanted to take out of the house where she was and marry her and have children, and he was asking if both of them could be cured of their disease. The doctor said, "Yes, you can both be cured, but it might be a terrible wrong to the children if you should have cohabitation until a year after the cure. Can you promise me that?" The couple hesitated and then promised. They were married. They met the doctor's condition. "The children came, three of them," said Dr. Rothstein. "I can say they were wellborn, and it has been for twenty years a successful marriage and still is." Perhaps one moral in the realm of the relations of men and women is: "You never can tell."

•   •   •

The case of Fred Burley and Jennie Bernberger was something else. The two of them had been going together in a rather easy way. They were teased more or less about how often they were seen wandering by themselves, keeping away from others. They were both fourteen, nothing special about either of them as to looks or cleverness, a very ordinary couple. The talk about them, however, was away out of the ordinary. One cool late-fall afternoon they were seen by three boys we knew who couldn't have twisted what they saw into something else than what they saw.

On the way we often took to the Seventh Ward schoolhouse we passed the roundhouse of the Fulton County Narrow Gauge Railway. It was a small affair, this railway, running some sixty miles from Galesburg to Lewiston, Illinois. The roundhouse was a small affair too, at times with no locomotive there and no one to attend a locomotive. The round table where the locomotive was turned around so it wouldn't run backward on the next run was seldom used and had a lonesome look. Under the rails and beams of the turntable, weeds and grass grew thick and there were many odd corners. We had played hide-and-seek and follow-the-leader there. The place said, "I will hide you" to any who might want to be hidden.

The three boys passing by took a notion to run out and chase each other along the beams. Suddenly in the grass and weeds of an odd corner two boys saw Fred Burley and Jennie Bernberger. They motioned the third boy to come look. They watched a minute or two and then ran away. They told it to the gang. One or two told it at home. It became one of those quiet scandals that don't spread far. Fred Burley's father was a railroad engineer of good record who liked his liquor. Jennie Bernberger's father was a Q.* machinist and they were known as a quiet decent family. Fred was neither handsome, clever, nor reckless in any unusual way. He was a good ballplayer and in a fight took pretty good care of himself, having an older brother who had been brutal in two schoolyard fights. It

*Chicago, Burlington & Quincy Railroad.

seemed hot blood ran in the family. What had happened, said some of those who talked about it, was one of those things that happen and you can never be sure why they happen. Someone got word to the Burley and Bernberger parents, and Jennie and Fred were never seen walking together again. In the course of time it passed away as nothing to talk about. No baby came.

Everybody passed it over as a little affair to forget — except one Swede boy. Of the five or six other boys I knew who talked about it while it was still a live and warm scandal, none of them mentioned it to me in the years that came after. This one Swede boy, however, was haunted by it. He had a mind and temperament for religion. He was a mystic and over his manhood years became a cultist in weird Christian sects and doctrines, writing tracts and speaking of himself, by inference, as a purist and a perfectionist. He carried himself as one who has attained holiness. I met him occasionally at intervals of ten and twenty years, and each time in our talk he brought up the sin of Fred Burley and Jennie Bernberger under the Narrow Gauge turntable long years ago. It was more than forty years after that boy and girl fell into their act of folly and passion that my religious friend wrote a letter to me and went out of his way to write his horror of those childhood sinners. He asked if I remembered the affair, with the implication that if I wasn't haunted by it as he still was, then I should properly be. I haven't yet figured out what it did to him and why, as with nobody else, it never changed for him in its color of blood scarlet. After forty years he shrank from the shame of their loathsome and evil act.

I am sure that if I had been struck as deep as he was by the infamy of this one event so memorable to him, I would go on a search. I would seek all items of information that would throw light on what happened afterward to Fred and Jennie. Did they marry, and how did their marriages turn out? Are there children, and how have they turned out? What does either of them, now in their seventies, have to say about one sudden event away and far back in the storms of their lives which they perhaps remember only dimly? Could one or both of them have gone on farther in sin so that if my friend should meet and hear them speak frankly they would paralyze him with what they know of the ways of evil? That could be. Or again they might tell of plain humdrum lives and perhaps one or two sons with Distinguished Service Medals or Purple Hearts. If we could know the facts we wouldn't do any imagining. We can make one inference to the effect that a zeal for holiness can be carried so far that it is both pathetic and comic. Unless this should meet the eye of my old Swede-boy playmate I expect next time I meet him he will again mention with pious aversion the piteous and unholy affair of a boy and girl under that Narrow Gauge turntable — sixty years ago!

Quite different was the atmosphere of another affair tinted with scarlet. Where I worked in the pottery at the southeast corner of town I often saw the young man Harry Wilson. He was a molder. He threw a lump of clay into a plaster-of-Paris mold of crock shape, put the mold on a turning wheel, kept the clay moist with the proper amount of water his hands kept throwing on it, and held a scraper taking the excess clay away. It was fun to watch him. I used to go when the chance came from the ground floor where I worked to the second-floor molding room. There I would see the crocks taking shape while I listened to the fascinating singing of Harry Wilson. Aye, the lad had tunes to him. Handsome he was, with a clean-cut face, straight nose and chin to match, liquid brown eyes, wavy brown hair, and a mouth for carving song words. Like his mother back in Ohio, he was an Irish Catholic. He had from her some old Irish jigtime ditties and come-all-yes. These he could sing, and shift to "Kathleen Mavourneen," "In the Gloaming," "The Last Rose of Summer," and other sad and poignant lyrics. My favorite was "Juanita." The fountain, the summer moon, the parting lovers in their last kisses, they came alive. He believed in love and its hungers and could give them in shadowed tones. I, a kid apprentice, had the nerve to say, "You ought to be with Al Field's minstrels." He smiled, "Just a molder. Once a molder always a molder."

The girls in the molding room carried the moist clay molds to a kiln for heat, drying, baking. One of the girls, Hilda Ellquist, of Swedish Lutheran peasant parents, was slow-minded, of robust build, with a comely face and a sweetness of clean strengths. You might see her in some of the paintings and sketches of Jean François Millet. I saw a rare dreaminess come over her face when listening to Harry Wilson sing "Juanita" or "In the Gloaming." The days arrived when Hilda wasn't coming to work. After a week or two, a girl who had seen Hilda reported with a sorry face, "She's in the family way." In this same week Harry Wilson didn't show up for work. The word was, "He's skipped town." And no report ever came back on where Harry Wilson had gone and what might have become of him and his voice and the further girls and women he met and sang for.

Hilda's baby came wellborn. She had an elder sister who was a perfect companion and comforter. Hilda carried herself with a quiet dignity, with neither pride nor shame. She worked as a housemaid, took a job in a laundry, kept on going to church, and brought along her boy with care and affection. Her father, a Q. day laborer, and her mother joined in care and affection for the little one.

A young farmer came along and began "keeping company" with Hilda. He played with the growing boy and had fun. He said to Hilda, "Why shouldn't we have a boy or girl of our own?" "You mean," asked Hilda, "we should get married?" "Sure — why not?" he said and he wasn't laughing. So they were married and Hilda bore him four bouncing babies, "thumping bantlings," worth looking at.

And her firstborn child? The years ran fast. He went to school, he went to war, he went into the selling game, a hardware line. A woman who saw him at a party reported to me, "He's made a success. You can see it the way people take him. He's a Presbyterian, an Elk, a Moose, a Mason, and is active in the Veterans of Foreign Wars. He saw combat duty in France in 1918. One of his three boys when last heard from was flying a helicopter in Korea." I had to say, "At this party did they ask him to sing?" "Not that I noticed." "Did any of them mention him as a singer?" "Not that I heard. Why do you ask?" And I passed it off with no word about a fellow she had never heard of, that sweet-singing bad-boy Harry Wilson, gone where the wind listeth, singing of love and its hungers.

There were a few folks who raised their hands as though to brush away evil, and they as much as called Hilda a slut and a strumpet. Still others laughed at the mention of Hilda's trouble but their laughter was as though someone had taken a fall on a slippery place — and no telling who's going to be next. I can't remember among the young people I knew that there was any of the spirit of Mr. Hawthorne's novel The Scarlet Letter, which was required reading in the high schools then. I missed going to high school but I read Mary's copy. I recall another boy who had got hold of The Scarlet Letter and how he told us, "It's about a preacher who knocked up a woman and a baby came and there was hell to pay. It got worse and worse, everybody fussed up. I quit reading it. I didn't care how it ended." He said more and I got the impression that the book not only told him that sex is a dangerous fire to fool with at any time, full of secrets and cruelties, but worse yet, sex is a sort of filth and flesh is nasty. Could it be that The Scarlet Letter reeks with Sex Education?

• • •

One brave woman I saw almost incredibly loyal to her marriage vows and family was the wife of a Q. shop laborer. Six healthy children had come to them. The man was known to the grocers as "good pay." He could buy "on tick" and no questions asked. They were among the families classified as clean and decent. Then the husband and father came down with consumption. They were living in the little house on South Street where our family had moved from my birthplace. I was delivering milk to them. The time came when their bill for milk tickets had run so high that my boss ordered me to let them have no more tickets. The sick man lingered, grew pale and thin till he seemed a bag of bones. Over two years he lingered, holding to life with white skeleton fingers. During that long struggle to live and make a comeback his wife "took in washing." I saw her at the tub many a time. She was the support for a family of eight in a three-room house. Day after day her hands ran along the washboard, ran clothes through a wringer, hung out the clothes and ironed them, one of her boys hauling the finished work on a little wagon to the customers.

She was not a large rugged woman who could take such work in her stride. She was small and well muscled, but the endless monotony of one wash after another and the same nice lace every week from one family and the same red-flannel underwear from another — it wore her till she was pale and thin. Friends and neighbors wondered if she would go down like her husband. But she didn't. Something in the pith of her bones told her to stand up and go on. This invisible something might have been devotion to her husband and little ones. It might have been some code in her blood from her Swedish ancestral stream, telling her life is a battle and you fight to the last with whatever you've got. In what she did there was love of her man and of her brood of those born from her — and there was also pluck and valor as strictly as that for which genuine heroes are awarded medals of gold with rich inscriptions.

She saw her husband die in the house where she had endlessly attended him and where she and the children had for nearly two years borne their burden of pity at the spells of racking coughs and the toils of keeping floors and garments clean. The widow slowed down somewhat in her battling at the washtubs, brought her boys along to where their earnings helped and she could rest her bones and look back on harder days and worse nights. I heard of her boys and girls on different jobs, decent and honest. I heard her say once, "I did my best with what I had. Sometimes I nearly give out, but I went on. My husband was a good man. The children have come out pretty well. They were such handy little helpers. All we can say is we did the best was in us." A spare and a gaunt life she had. A citation could be written over her grave: "She was in her humble pathway a heroine who illustrated the meaning of true love."

There was a stanch little woman, strong-bodied with a rich full bosom, whose husband thought the world of her, whose neighbor women spoke admiration of how neat and handy she was with her housework. She could fix a pie for the oven or do a washing with the least number of motions needed, singing at her work. And her neighbors knew she often whispered a prayer about a deep wish she had. When her first child came it lived only two days. Her second child lived only a day, and I heard of two neighbor women who talked to each other about it and suddenly the two of them had their aprons to their eyes drying the tears. I heard a boy say that his mother had gone over to help in any way she could. She had seen the dead child and there came from the mother the half-choked words, "I think God ought to entitle me to one child, don't you think so?"

When Hattie Hoffenderfer left our block and her home and her folks we wondered a little about how she would do in Chicago. "She's got nice hips and a good bust, but her face ain't nothin extra," I remember one boy saying. In a year or two we had forgotten Hattie. Then news came dribbling back that in one of the big department stores she had moved up from clerking to an important office job. Another year or two and we had again completely forgotten Hattie. Then one summer she came back to the home town, on a two weeks' vacation, visiting with her folks and looking up old acquaintances. The word went around, "Have you seen Hattie Hoffenderfer?" And some who had seen her said, "God, the clothes she's got on!" She was the flashiest-dressed woman who had ever walked around our end of town. She had the tightest-fitting corset and the highest-heeled shoes we had ever seen. Her walk was a continuous strut. When she talked with old neighbors and friends she didn't overdo it but she let it be known that she had come up in the world to an important position. What she held down was a "position" rather than a job.

The silk she wore looked genuine and the three gold rings on her fingers, set with jewels, and the diamond brooch at her throat. Had she talked more about her work and what she did to earn the money to buy such apparel and adornments, there wouldn't have been the kind of talk that went around. She liked Chicago. "Chicago's been good to me," we heard her say, and we answered, "It sure has." No one we knew of dared to ask her, "Hattie, where in all time did you get the fancy duds and stones you're wearing?" We doubted that when she took the train back to Chicago she had the least notion of what kind of gossip buzzed around her. The slang of the day had it, "She certainly could put on the agony." How her German peasant father and mother took her visit I didn't hear. The father kept on going regularly to his job in the Q. shops.

● ● ●

Farnham Street where it meets Fifth was a country road when I walked barefoot in its dust, the land pasture or cornfield. On a piece of ground a house went up there and in that house Galesburg had the bloodiest crime of passion in its history. The man had been drinking but he couldn't have been clumsy drunk, for the fast and horrible work he did on the night he came home to find a man in bed with his wife. He sent a bullet into the man's body and a bullet into his wife's right arm. He put two razor cuts in the palm of his wife's right hand and slashed a deep cut in the left side of her neck from below the ear to the mid-line of her throat. Across the man's throat his razor swept from the left ear nearly to the right ear, severing the jugular vein. The bed was blood-soaked and the pools not yet settled when police officers arrived in the morning.

"The fight lasted two or three minutes," the killer testified when on trial. His movements around the unlighted upstairs room must have been fast, his feet quick and his hands working in a wild fury. He was found guilty of manslaughter and sentenced to fourteen years in the state penitentiary. Mart joined eighteen character witnesses — among them a college professor, a physician, and men the killer had worked for — who testified that "before his recent trouble his general reputation as a peaceable and law-abiding citizen was good." He served his time in Joliet, came back to Galesburg, lived a quiet life, and died, having had occasional news from his two married daughters and grandchildren in Chicago. Mart said to me, "He was that gentle and easy-going that he was about the last man in town I would have expected to hear was a killer and a double one. He told me that once before the last awful night he had caught the same man in bed with his wife. He warned them but they stayed in bed and jeered at him and said he didn't have the nerve to kill a flea." After he had served five years, Mart and others tried for a pardon or a parole but couldn't swing it. The affair was involved, complicated in motives, and not lacking in light on why hard liquor in excess is termed "tanglefoot."

As a boy I met unfinished stories, small scraps of everyday life not rounded out. On a streetcar to Highland Park, one of two men in the seat in front of me raised his voice at one point to say, "My wife doesn't like it and what she doesn't like I just naturally have to hate." I know sophisticates of this later age who would remark, "There seemed to be a maladjustment between them."

I heard a boy telling, as he heard it from his mother, of the woman next door to them. "She found a loaded revolver he had brought home and hid. She took out the cartridges, afraid he was going to kill himself or her. She didn't tell him about takin out the cartridges. A week later she found the revolver again and it was loaded and she took out the cartridges and didn't tell him about it. It's a kind of a game they play now. He don't tell her about loadin the revolver and she don't tell him about unloadin it and my mother wonders if somebody is goin to be shot. She wonders if she could give advice that would be any use. She's gettin nearly as nervous as the woman next door."

There was a house where a railroad engineer lived. He had moved into that house with two children and had seen six more children arrive there. He had made good money and most of his children finished high school. One child died early of diphtheria, another met an accident and would walk lame through life, two of the girls married men who drank hard and ran with other women. Only two of his children turned out somewhat as he would have liked. I happened to be selling the weekly Sporting News in the Q. depot and overheard the father saying to another engineer that he was waiting for a train to bring in one of his boys who had done fairly well. And I can't forget the offhand yet positive way that he remarked, "Eight children is too many."

Walking a few feet ahead of me on Main Street one day were two railroad men, young brakemen. An interesting brunette in a wide picture hat passed and one of the men turned his head for another look at her. Then as they walked along he said, "I don't think any man should stop lookin!" And after a pause he gave his second thought, "If my wife heard me say that she'd raise hell."

On a warm Saturday evening one summer on the Public Square near the Union Hotel I saw two young sports flashily dressed, a little flush with liquor, and I could tell by their voices they were in a quarrel or a hot argument. As I passed what I heard stayed with me, one of them holding the other by the coat lapel and snarling, "I know you better than you know yourself and I'm tellin you you're a damned queer duck and you're goin to have a hell of a time when you get tied up with any woman. You'll never get what you want and if you do it won't last."

I heard a high-school girl tell of her "strict" mother worrying about her. A boy with a name for being wild with both boys and girls had taken this girl for a long walk one night. She told her mother that they walked as far as the Lombard campus and she refused to go strolling among the pine clumps or the tall grass with him. Then came the questions: "Did he get fresh with you?" "How do you mean?" "Did he touch you, did he lay a hand on you?" "No, but I wouldn't care if he did." What came about later in that house between mother and daughter I didn't hear but I am sure there was a conflict of wills.

I was perhaps twelve years old when I got a job in haytime on a farm near town, twenty-five cents a day and a noon dinner. My job was carrying to the men in the field a jug of cold water, wet gunny sack wrapped around the jug to help keep the water cool. They were rougher men than I had ever worked with. I couldn't follow all their jokes and brags about women they had been with. At the dinner table, after their first attacks on the food had broken their hunger the three men talked and laughed. Suddenly Mr. M., the man of the house, the tenant farmer who was our boss, went into a blaze of anger and snarled a mean remark to his wife, who kept the house and had cooked and served our dinner. I have forgotten what he said and what he was blaming her for. At our house I hadn't seen or heard the like. The woman stood still. The men were quiet and we all looked at the woman. We saw she didn't shake nor choke. She just stood still and cried without making a sound. The tears ran down her face and she lifted her apron and wiped her eyes and cheeks. Mr. M. mumbled and muttered, like storm thunder dying away. We took up our eating again and there was no more talk and laughing. The men went to the field and stuck their pitchforks into the haycocks. The man on the hayrack pitched and shuffled it into a nice oblong stack and I brought them the jug they lifted and tilted so the cool water ran down their throats. And over and over I would wonder about the woman at the house and her crying and how many times he had snarled at her and slapped her face and done worse, maybe hitting her with his fist. What I caught was that he got some kind of enjoyment out of snarling at her and shaming her before other people. I couldn't make out what was going on, only whatever was wrong I was on her side and against him. After three days the hay was in. I was paid off and never heard what became of Mr. M. and his wife that he mistreated and before other people.

What was there so strange and sad when it first came to your ears about a man and his wife, "There's no love lost between them"? You worked on it and it came out that they had found a love between them and then lost it.

•   •   •

Roland Worth was a college athlete, a fine quarterback, with black hair and black eyes, bright in his classes. His father held one of the most important and responsible positions in the city. They were well off as to money. Mrs. Worth many a time held out a crock for a quart of milk I poured. I would have believed that her son could have had any girl in Galesburg he wanted. It was years later that I talked with a trustee of Knox College, a beautiful woman whom I had adored at a distance in my milk-wagon days. She gave me a conversation she had with Mrs. Worth, who opened it, "You ought to marry my son." "Why, he's never told me that he loves me." "He does love you, he worships you." "If anyone loves me I expect him to tell me so again and again, then I'll believe him." So there in the house of one of my milk customers was the play of <u>Love's Labour's Lost</u>.

"Why don't you pick you a nice young woman and marry her and settle down?" was asked of a slick young passenger brakeman, who answered, "Why should a man stick with one woman when he can satisfy all of them?" He ran wild for a while. Then he did meet a girl he couldn't get away from, married her and settled down. Two children came. He was promoted to conductor. Two more children came. And what you heard about him was, "He's dependable every way you take him."

Later a Peoria lawyer told me about one of his cases, a suit for divorce, the husband charging adultery. The leading witness worked as a chore man at the house where it was alleged the offense was repeatedly committed. He testified that on a certain afternoon of a day he named he suddenly remembered a piece of repair work on the front porch of the house. He had put it off for weeks and decided to get the job done on this afternoon. At the end of the porch, before starting work he happened to take a look through the window of a bedroom. He was asked by the husband's attorney what he saw. He answered that he saw the accused woman in bed with a man. Then came the question, "What did you see them doing?" His answer came with no hesitation, "What did I see them doing? Well, there they was, toes to toes!"

A Galesburg man had a visit with a Chicago relative of his who was in Joliet for life for killing a woman and was saying, "There's lots of time to think here. The only thing I miss in the pen is women." He hesitated, and then, "But if I was outside I might meet another woman like the one I killed."

There was a period with my playmates, mostly Swedish boys, when going to and from the Seventh Ward school, we had ideas or little frames on which we tried to weave ideas while lacking the skeins of experience. We were along ten or twelve years of age. When we talked about our families and what class of society we were in, we said, "We ain't rich and we ain't poor." As a statement it would hold good. We were on that boundary line where so definitely we needed ten thousand times what we owned as property in order to be rich, wealthy or affluent, and a drop of thirty to fifty per cent in wages would have put us strictly into the classification of the poor, the genuinely poverty-sunken. Those who have lived on that delicate, shifting, and hazardous boundary line have had an experience that some use as help in understanding economic issues and the labor history underlying the swirling human currents of our times.

At this time we ten- to twelve-year-olds had positive views on marriage. We were against the institution and practice of holy matrimony. Most of us were sure we would never go in for marriage. We had seen enough of it. I recall how bluntly one boy put it, and the rest of us chimed in and agreed with him: "If you marry you have to sleep in the same bed with the woman. Then the babies come. And you have to raise children and they eat and get sick and wear out clothes and take up room and make a mess around the house and they cost more than they are worth." Each of us was sure he would somehow get along without getting tied to some woman and a house full of children that cost money and the man has to earn the money. In later days the question didn't seem to come up whether we were rich or poor and we seemed to move into a viewpoint that if a man found him the right kind of a woman it might be worth the time to marry though it was still a gamble.

A boy in our block I knew well, and I saw his love for a girl in the next block. They walked and held hands and life made rainbows for them. She had him do most of the talking. She was a slim figure, stunted in growth but shapely, and had an ivory-pale beautiful face. At fourteen they began "going together" and kept "steady company" for two years, saying they would marry at eighteen. Then came what was called "galloping consumption," lung hemorrhages, and for six weeks he went nearly every day to see her white-faced smile and to hold her thin little white hands. When Madeline Clark died he went to the funeral and came away and said to me, "Be a long time before I can look at another girl." He was years getting over it. Sometimes what the superior elders call "puppy love" runs deep and leaves scars.

I had my "puppy love." Day and night her face would be floating in my mind. I liked to practice at calling up her face as I had last seen it. Her folks lived in the Sixth Ward on Academy Street next to the Burlington tracks of the Q. They usually left a crock on the porch with a quart ticket in it. I took the ticket out of the crock, tilted my can and poured milk into my quart measure, and then poured it into the crock, well aware she was sometimes at the kitchen window watching my performance, ducking away if I looked toward the window. Two or three times a week, however, the crock wasn't there and I would call "Milk!" in my best boy-baritone and she would come out with the crock in her hands and a smile on her face. At first she would merely say "Quart" and I would pour the quart, take my can, and walk away. But I learned that if I spoke as smooth and pleasant a "Good morning" as I could, then she would speak me a "Good morning" that was like a blessing to be remembered. I learned too that if I could stumble out the words, "It's a nice day" or "It's a cold wind blowing" she would say a pert "Yes, it is" and I would go away wondering how I would ever get around to a one- or two-minute conversation with her.

I was more bashful than she. If she had been in the slightest as smitten as I was, she would have "talked an arm off me." But she didn't. It was a lost love from the start. I was smitten and she wasn't. And her face went on haunting me. Today I can call up her girl face and say it's as fine as any you'd like to rest your eyes on, classic as Mona Lisa and a better-rounded rosy mouth. I had no regrets she had smitten me and haunted me. I asked for nothing and she promised the same. I could say I had known my first love. It was a lost love but I had had it. It began to glimmer away after my first and only walk with her.

I dropped in with another boy one summer night to revival services at the Knox Street Congregational Church. There I saw her with another girl. After the services a chum of mine took the other girl and I found myself walking with the girl of my dreams. I had said, "See you home?" and she had said, "Certainly." And there we were walking in a moonlight summer night and it was fourteen blocks to her home. I knew it was my first or last chance with her. I said it was a mighty fine moonlight night. She said "Yes" and we walked a block saying nothing. I said it was quite a spell of hot weather we had been having. She said "Yes" and we walked another block. I said one of the solo singers at the church did pretty good. And again she agreed and we walked on without a word. I spoke of loose boards in the wooden sidewalk of the next block and how we would watch our step, which we did.

I had my right hand holding her left arm just above the elbow, which I had heard and seen was the proper way to take a girl home. And my arm got bashful. For blocks I believed maybe she didn't like my arm there and the way I was holding it. After a few blocks it was like I had a sore wooden arm that I ought to take off and have some peace about it. Yet I held on. If I let go I would have to explain and I couldn't think of an explanation. Not for a flickering split second did it dawn on me to say, "You know I'm crazy about you and crazy is the right word." I could have broken one of the two blocks we walked without a word by saying, "Would you believe it, your face keeps coming back to me when I'm away from you — all the time it keeps coming back as the most wonderful face my eyes ever met." Instead I asked her how her father, who was a freight-train conductor on the Q., liked being a conductor and did he find it nice work.

We made the grade at last. The fourteen blocks came to an end. I could no more have kissed her at the gate of her house than a man could spit against Niagara Falls and stop the water coming down. Instead of trying to kiss her I let go her arm and said "Good night" and walked away fast as if I had an errand somewhere. I didn't even stand still to see if she made it to the front door. I had made the decision I wasn't for her nor she for me. We were not good company for each other. If we were, at least one of us would have said something about what good company we were. I still adored her face and its genuine loveliness, but it had come clear to me that we were not "cut out for each other." I had one satisfaction as I walked home. My bashful right arm gradually became less wooden. The blood began circulating in it and my fingers were loose instead of tight and I could wiggle them.

I have an album of faces in my memory, faces that were a comfort. One I saw many times. She would be walking a street or driving a surrey. Her face was a blossom in night rain. She married a small storekeeper and children came. And her face was still to me a blossom in night rain. I never spoke a word to her. What I adored about her might have been mostly imagination. Yet I heard long after from those who knew her that she had an inside loveliness matching her face.

I remember several beautiful girls and women of my home town. Their personal loveliness compares nicely with that of notable stage stars and society women. I remember an Irish girl with a turned-up nose, flashing brown eyes, and a swift impudent mouth. She had her kind of beauty, as did a Swedish girl whose slow smile had a prayer in it while her walk was the water willow moving to a morning wind. There was the singer heard on many local occasions, her contralto voice and personality lovable. She was the wife of a railroad fireman and there were gossips malicious in their talk about her. There was Mary's chum of many years, Anna Ersfeld, as stunning a brunette as any in the Floradora Sextette. There was Anna Hoover, the public librarian, with a rich bosom and an apple-blossom face framed in dark hair. There was Janet Grieg, the young matron of the Knox girls' dormitory, Whiting Hall, Miss Grieg a mezzotint by Copley, a country girl from a farm near Oneida teaching manners to the Knox country girls. There was fair-haired Ollie Linn and bonnie Lois Smith, the robust and rosy Rilla Meeker, the gay singing Nell Townsend, Alice Harshbarger, whose head would have been welcomed by any painter seeking a model, Frankie Sheridan with a face of Maytime bloom. And so many times Mary spoke words of faith that counted with me.

Over in the Fifth Ward near St. Patrick's Church of a morning on the milk route I would meet Theresa Anawalt starting to her job in a Main Street store. My eyes would be on her walk and the ways of her head and shoulders long before we met and passed by each other. As she came closer, my eyes fed on the loveliness of her face. I'm sure her face is there in certain Irish ballads of wild fighting over such a face. A hundred times I met her about eight o'clock of a morning coming along the same sidewalk in the same block. She seemed to have thoughts that were far away and over the hills. About the fiftieth time we met I figured it might be time for me to say to her with a bright smile, "Well, here we are again." And then, figuring some more, I believed I would leave it to her for such a remark. It wasn't with me a case of love at first sight or the hundredth sight. It just happened that I found her wonderful to look at, a mysteriously beautiful young woman with a sad and strange mouth. We didn't speak to each other there and then on that sidewalk in that block in the Fifth Ward — nor ever afterward. And I haven't heard her name or any scrap of news about her since that year — Theresa Anawalt. And if this stray item about her should meet her eye, she can't say I've forgotten her. I have in this chronicle, for the sake of convenience, given fictitious names to some persons, but not in the case of Theresa Anawalt.

I could go on. It would be quite a little gallery. I could say that I carried faces with me and could turn album pages looking at them when they were gone I didn't know where. I enjoyed their loveliness in my boy's mind in ways they could never have guessed. I once heard a man say, "The town I grew up in didn't have a woman worth a second look." Either his eyes were not so good or it was a hard town for him to grow up in. Of course, I could mention the drab and the tragic that came to some of my album women, but I knew them in their Springtime Years when a freshness of dawn was on them before time and fate put on the later marks.

On the Spread
of Venereal Disease

Shortly after World War II an article reprinted in the *Reader's Digest* asked whether the newly discovered miracle drugs were about to destroy the nation's morals. [In essence, the author asked why anyone should refrain from sexual adventures now that gonorrhea and syphilis can be cured by one or two painless injections.] Why indeed? The same question was asked by the Church in the 16th century when mercury was advanced as a treatment for syphilis. Even before the *Reader's Digest* arrived on the scene, disease was looked upon as an appropriate punishment for sin.

The early enthusiasm over the success of miracle drugs has given way to very sober second thoughts. Resistance to antibiotics on the part of all bacterial populations has grown hand in hand with the use of these drugs. Neither the bacterium that causes gonorrhea nor the spirochete that causes syphilis is an exception. Today's arsenal against venereal disease simply is not as powerful as it once was; the arsenal has remained basically unchanged, but the target organisms have evolved.

The incidence of syphilis in the United States was once exceedingly high. About 1 person in 15 was infected and in need of treatment during the late 1930's. By heroic public-health measures and educational campaigns this reservoir of disease was reduced nearly tenfold by 1957, a year in which only 6,000 new cases were reported. We are now losing the battle once more. About 20,000 new cases of syphilis have been reported annually since 1961. New cases of gonorrhea are not reported as accurately as are those of syphilis; the estimated incidence of new cases of gonorrhea is about 2,000,000 each year. As many as one American woman of every 100 may harbor gonococcus, the bacterium responsible for gonorrhea.

Many young persons look upon sexual experience as just that – a wholesome experience, a casual adventure, a release of pent-up tensions, or an expression of affection for another person. So far, so good. Even the young people of Galesburg had roughly the same ideas a century ago. The chief difference between the two groups of youngsters, then and now, is illustrated by the following dialogue that took place in countless homes late on countless evenings: "Where have you been?" "Out." "What did you do?" "Nothing." Children in the past were not verbal.

The person to whom the following calculations are addressed is the one who circulates – like the young lady mentioned in the introductory comments who garnered a dozen "character" references by seemingly spectacular abilities. We must assume that the young men she knows circulate in a similar manner, that each one of them from personal experience can offer references for a half-dozen or more other young ladies. We also assume that the entire system of interlocking liaisons is a reflection of overlapping circles of close aquaintances, much like the overlapping circles shown in the accompanying figure. The use of many circles in the construction of this diagram recognizes that there are favorite partners in the love game but also recognizes that enough additional exchange takes place to give the entire, irregular geometric shape an overall coherence. For convenience, I refer to the persons within the entire complex as an "association."

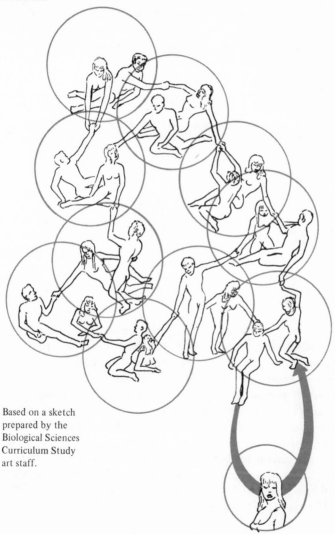

Based on a sketch
prepared by the
Biological Sciences
Curriculum Study
art staff.

Outside the interlocking association of the adventurous young people in our diagram is a source of infection, a reservoir of VD – the prostitutes of the neighborhood or other, previously infected associations of young persons. Suppose that there is one chance in a hundred that a fellow (we shall put the initial blame on the male) will step outside his association and contract VD from an outsider. What are the consequences of this constant threat of infection as far as the association is concerned? The answer lies in the number of couples involved in the interlocking association. In the accompanying table I have shown the probability of an introduction of venereal disease into the association as it changes with the number of couples involved. If the association consists solely of one couple that keeps to itself (except for the straying of the male as postulated above), the probability of infection per year is only one in a hundred; 95 of every 100 devoted couples of this sort could maintain their relationship for five years (throughout their undergraduate years, for example) without contracting VD.

Number of Couples in Group	Probability that Venereal Disease Will Be Introduced into the Association in a Given Year	Probability that the Association Will Escape Infection over a 5-Year Period
1	0.01	0.95
2	0.02	0.90
3	0.03	0.86
4	0.04	0.82
5	0.05	0.78
10	0.10	0.61
15	0.14	0.47
20	0.18	0.37
25	0.22	0.29
50	0.39	0.08
100	0.63	0.01

As the numbers of persons in the association grows, so does the probability of infection. I do not know how large the whole ramifying structure must be if one girl is intimate with a dozen lovers but I am inclined to think that the number of couples involved must be substantial. In this case it may be a toss-up whether the group can get through a single year without becoming infected under my postulated rate of infection from the outside reservoir. How rapidly the infection would then spread *within* the association is unknown but presumably, unless the boasts one hears are but empty ones, at a rather rapid pace. If an association tends to remain more of less undisturbed for years (as through high school and college), then the probability that VD will enter the group within a period of five years may be 50:50 with even a relatively small number (15 couples) of participants.

The calculations made here are not unrelated to those made in an earlier essay on the arithmetic of epidemics.* We are, you see, discussing what has become a truly disturbing epidemic of venereal disease within the United States. Freedom of experience for the individual and freedom to spread for the infective organism are about the same in sexual matters. The estimate of one male in every 100 becoming infected per year was, of course, picked for illustrative purposes. As one association after another becomes infected, however, the size of the reservoir of infected persons becomes larger, and so the probability of infection grows for errant males of still uninfected associations.

That the probability of infection by **VD** is least for single couples is no accident; it is from such couples that marriages grow. One can, indeed, look at marriage as a strategy for minimizing the probability of contracting venereal disease. The probability is reduced to nearly zero, of course, if both partners have postponed sex until marriage or, if not, if they have remained exclusively with one another. In many states, to the horror of those youngsters who are trying hard to be modern, the latter relationship could qualify as a common-law marriage — a concept whose roots lie in the 13th century.

*See the essay entitled "The Arithmetic of Epidemics" in this volume.

An Outbreak of Venereal
Disease in Massachusetts

The preceding essay was written as an intellectual exercise on the epidemiology of venereal disease. In the meantime I have encountered a study of the spread of venereal disease in a Massachusetts community during the fall and winter of 1955 and the following spring and summer.* The study is a small classic. In the following I have reproduced the cast of characters, a diagram illustrating their interrelations with one another, and extended excerpts from the original professional report. One comment from the original paper must be repeated here: "From their point of view, premarital and extramarital relations were considered natural, and were therefore practiced without a sense of guilt ... Already, reinfections have occurred among several members of the group, *particularly 'ping-pong' gonorrhea.*" (emphasis mine)

Cast of Characters
(In order of their appearance)

Alice
  A. Allan
  B. Boris
    1. Betty
    2. Charlotte
      2a. Charlie
      2b. Donald
        2b-i. Charlotte (see 2)
        2b-ii. Dorothy
        2b-iii.Olga
        2b-iv. Pamela
        2b-v. Ruby
        2b-vi. Sandra
      2c. Edward
    3. Dorothy (see 2b-ii)
      3a. Donald (see 2b)
      3b. Frank

      3c. Dennis
    4. Ellen
      4a. George
        4a-i. Frances
      4b. Harold
      4c. Irving
    5. Gilda
      5a. Jack
      5b. Kenneth
      5c. Larry
      5d. Mark
    6. Helen
      6a. Neal
      6b. Oscar
        6b-i. Janet
          6b-i(1) Peter (see 7b)
        6b-ii. Helen (see 6)

*N. J. Fiumara, J. D. Shinberg, Evelyn M. Byrne, and Jeanne Fountaine. "An Outbreak of Gonorrhea and Early Syphilis in Massachusetts," *The New England Journal of Medicine*, 256 (1957), 982-990.

6b-iii. Marion
    6b-iii(1).   Zachariah
    6b-iii(2).   Albert
6b-iv. Irma
6c.   Bob
   6c-i. Helen (see 6)
   6c-ii. Tina
   6c-iii. Ursula
   6c-iv. Carl

7. Irma
  7a.   Quentin
    7a-i. Karen
      7a-i(1). Peter (see 7b)

7a-i(2). Walter
7a-i(3). Xavier
7a-i(4). Sam
7a-i(5). Quentin (see 7a)
7a-ii. Lorraine
7b.   Peter
  7b-i. Janet (see 6b-i)
  7b-ii. Karen (see 7a-i)
7c.   Oscar (see 6b)
7d.   Robert
7e.   Sam
7f.   Tom
7g.   Ulysses
7h.   Victor

The Spread of Venereal Disease in a Local Community

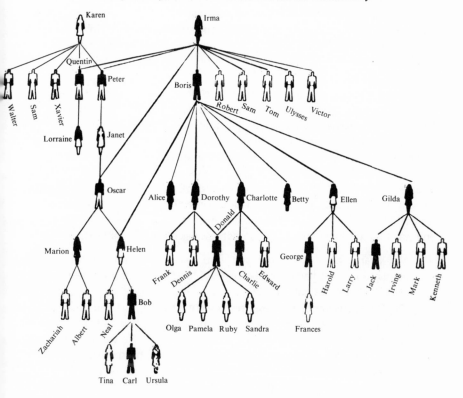

Descriptive Account
of the Epidemic

The case that furnished the first clue to the epidemic was that of Alice, an attractive nineteen-year-old married woman with a two-year-old son. She consulted her family physician because of a bad cold, which she had been unable to "shake off" for the past month. It was characterized by a sore throat of three weeks' duration, headache, loss of weight, low-grade fever, "glands" in the neck, weakness and lassitude. The fever ceased in ten days, and Alice did not recall having a rash. Physical examination revealed slight infection of the pharynx and cervical adenopathy. The family physician made a routine blood Hinton test, concluded that the patient had a severe upper-respiratory infection, which was prevalent that winter, and asked her to return in three or four days. She was treated symptomatically with aspirin. The report of the blood Hinton test was positive.

Alice worked in a shoe factory to support herself and her child, even though she was married. Her husband seldom worked and changed jobs frequently. He was a heavy drinker, got drunk frequently, and when under the influence of alcohol beat his wife unmercifully. Because of this she had been separated from him for varying periods, but she always took him back.

The family physician, knowing a little of Alice's social background, referred her to the State Cooperating Venereal Disease Clinic, where a diagnosis of early latent syphilis was established on December 23. Treatment was started on that date. The nurse venereal-disease epidemiologist interviewed the patient and was told that the only sexual contact was her husband (Table 2).

A. Allan, the two-year-old son, was examined although his mother's prenatal blood test had been negative. A blood Hinton test was negative, and he was discharged as not infected.

B. Boris, the husband, was twenty-eight years of age and good-looking, with dark, curly hair. When his wife told him that she had visited the clinic he beat her and was arrested, spending the night in jail. He was brought to the clinic on the following morning by a policeman. The nurse epidemiologist found him very unpredictable, being good-natured and polite at times but surly and rude at others. Although he had abused his wife, he spoke highly of her and resented any intimation that she might have had an extramarital exposure.

Boris admitted that he had penile sores in October, which had disappeared. Blood tests were repeatedly positive; a spinal-fluid Wassermann test was negative. A diagnosis of early latent syphilis was thereupon established, and treatment was begun on December 22. Boris gave the names of seven contacts, as follows:

1. Betty, sixteen years old and single, had an exposure with Boris in October and repeated exposures in January and February. She was extremely attractive and worked only when she felt like it. She frequented the bars and taverns. The mother and father were separated. The mother was alcoholic. Betty had a younger sister (Helen, contact 6, below), who was also promiscuous, and an older brother who lived in another state. She believed that if she lived with this brother and his wife she would lead a better life.

Betty had a positive blood Hinton and a negative spinal-fluid Wassermann test. On February 17 the diagnosis of early latent syphilis was made, and treatment was begun. She continued to have exposures with Boris, and in July she was found to be four months pregnant. Her last period was on March 5. She denied any contacts other than with Boris.

2. Charlotte, a twenty-year-old married woman, visited her cousin (Dorothy, contact 3) in the middle of December and spent the weekend in town. The cousin introduced her to Boris, with whom she had repeated exposures over this weekend. She returned home, and a diagnosis of primary, ? secondary syphilis was made by her family physician on April 4, when treatment was begun. On interview she named 3 contacts:

2a. Charlie, the twenty-six-year-old husband of Charlotte, went to the family physician with his wife on April 4 becuase of a penile sore. The diagnosis of primary syphilis was made, and treatment was begun. He denied any extramarital exposures, but when his wife told him of hers he "bawled her out" and told her, "Don't do it again." We suspected that he had been having extramarital relations, since he did not act like an aggrieved husband.

2b. Donald, thirty-four years old and married to Olga (contact 2b-iii), worked as a bartender at a bar frequented by the patients and contacts involved in the epidemic. He admitted an exposure with Charlotte in March. The blood Hinton test was positive, and the spinal-fluid serologic test was negative. A diagnosis of early latent syphilis was established on May 1, when treatment was begun. He gave us the names of 6 contacts, only 1 of whom (Dorothy) was infected:

2b-i. Charlotte (contact 2). Exposure with Donald in March.
2b-ii. Dorothy (contact 3) had an exposure with Donald in December.

2b-iii. Olga, the thirty-year-old wife of Donald, was negative for both gonorrhea and syphilis.

2b-iv. Pamela, twenty-five years old and single, had exposures with Donald in November or December. She was not infected.

2b-v. Ruby, twenty-nine years old and single, had an exposure with Donald in March. She was not infected.

2b-vi. Sandra, twenty-one years old and separated from her husband, had contacts with Donald during November and December. She was not infected.

2c. Edward, a chubby, twenty-five-year-old single man, lived outside the State but worked in this community. He met Charlotte at the bar where her cousin Dorothy was employed as a barmaid. Exposure occurred in January or February in Edward's car. He was examined by his family physician, and the blood tests were negative.

3. Dorothy, the barmaid mentioned above, a twenty-eight-year-old divorcee, had remarried (to Dennis, contact 3c). She had introduced her cousin Charlotte to Boris. She herself had exposures with him during November and December. She was short, light-complexioned and a neat dresser, and looked like an old-fashioned "schoolmarm," wearing glasses with a black cord, a white blouse with collar, and a black bow tie. She usually had on a dark skirt and made a most dignified appearance.

On December 28, when Dorothy reported to our clinic, she was found to have secondary syphilis, and treatment was begun. She named 3 contacts besides Boris:

3a. Donald (contact 2b), besides his exposure with Charlotte, had one with Dorothy during December.

3b. Frank, twenty-two years old and single, had an exposure with Dorothy in December. He had had repeated examinations, including blood tests for syphilis. All these examinations were negative, and on June 25 he was discharged as not infected.

3c. Dennis was the twenty-six-year-old husband of Dorothy. He was not infected.

4. Ellen, a pretty but vacant-looking eighteen-year-old unmarried girl, had an illegitimate eight-month-old daughter. She lived with her married sister. Her mother was dead and her father lived outside the State. According to her sister, Ellen had always been a problem. She admitted an exposure with Boris on December 15. She was found to have gonorrhea (positive smears and cultures) and was treated for it on February 28, and was found to have early latent syphilis on March 13, when antisyphilitic treatment was begun. She gave us the names of 3 contacts in addition to Boris.

4a. George was twenty-eight years old, tall, dark and good-looking, a neat dresser with an aristocratic air. All the girls "swooned over me, and the men were jealous of me." He had exposures with Ellen in January. He did not want to be treated at the clinic but could not afford a private physician. He was finally examined at the clinic, and on May 4 the diagnosis of secondary syphilis was made. Treatment was begun on that date. In addition to Ellen George named 1 contact:

4a-i. Frances was his twenty-five-year-old wife. Repeated blood tests were negative.

4b. Harold, twenty-four years old and unmarried, had exposures with Ellen in December. Repeated examinations and blood tests were negative, and he was discharged as not infected.

4c. Irving, twenty-two years old and single, had exposures with Ellen during February and March. Repeated examinations were negative, and he was discharged as not infected.

5. Gilda, a nineteen-year-old woman, had been separated from her husband for two months. She had an exposure with Boris in December. She was not located until May, and on June 4 a diagnosis of early latent syphilis was established. She was then treated. She admitted 4 contacts in addition to Boris:

5a. Jack, a twenty-two-year-old single man, was a friend of Boris, who introduced him to Gilda. Exposures occurred during January and March. Jack had repeated positive blood tests for syphilis and negative spinal-fluid Wassermann tests. On May 1 the diagnosis of early latent syphilis was made, and treatment was begun.

5b. Kenneth, twenty-two years old and single, had exposures with Gilda during April and May. Repeated examinations, including blood tests for syphilis, were negative, and he was discharged as not infected.

5c. Larry, twenty-one years old, was Gilda's husband. He admitted a single exposure with her in February. Repeated blood tests for syphilis were negative, and in June he was discharged as not infected.

5d. Mark, a twenty-four-year-old man who was separated from his wife, admitted exposures with Gilda in April. Repeated blood tests for syphilis were negative, and he was discharged as not infected.

6. Helen was the fourteen-year-old sister of Betty (contact 1). She was attractive and looked older than her age. She was a product of a broken home, with the mother and father separated and the mother alcoholic. Her older brother (described above under contact 1) was the only stable member of the family. She had been in trouble with the police because she was a habitual truant and a stubborn child, and was under the care of the Youth Service Board.

Helen admitted exposures with Boris in December. She was found to be infected with gonorrhea on February 14 (smears and cultures positive) and with early latent syphilis on February 17, on which date she was treated for both infections. She gave the names of 3 contacts in addition to Boris:

6a. Neal, twenty-nine years old, was separated from his wife. He was short, with a medium-dark complexion, and looked about forty years of age. He met Helen at a bar and took her to the city dump, where exposure occurred on March 9. She remembered this date because it was the day of her last penicillin injection. Neal had repeated negative blood tests for syphilis and was discharged on June 8 as not infected.

6b. Oscar was a thirty-year-old man with three children, separated from his wife because "she was frigid." He then "fell in love with a tramp" (Janet, contact 6b-i). His being in love did not preclude his having exposures with Helen during December. On admission to the clinic Oscar had a profuse urethral discharge and a genital lesion that was dark-field positive. A diagnosis of gonorrhea and primary syphilis was therefore established on February 3, when treatment was begun. In addition to Helen, Oscar gave us the names of 3 contacts:

6b-i. Janet was the "tramp" with whom Oscar fell in love. He had repeated exposures with her from September through January 28. Janet was found to be infected with gonorrhea and was treated on February 2. She did not have syphilis. She named 1 contact besides Oscar:

6b-i (1) Peter (contact 7b). Exposure in January.

6b-ii. Helen (contact 6) had an exposure with Oscar early in December.

6b-iii. Marion was a thirteen-year-old girl, an eighth-grade student. She was known to the police authorities for her habitual truancy and was under the care of the Youth Service Board. Oscar had an exposure with her on January 20, and on February 25 she was found to have early latent syphilis and was treated. She named 2 contacts besides Oscar:

6b-iii (1). Zachariah, twenty-two years old and divorced, had exposures with Marion in December or January. He was not infected with either gonorrhea or syphilis.

6b-iii (2). Albert, twenty-one years old and single, had an exposure with Marion in January. He was not infected with either gonorrhea or syphilis.

6b-iv. Irma (see contact 7). Exposure in December.

6c. Bob, twenty-six years old and single, lived with his widowed mother. He had no steady employment and worked at odd jobs with an amusement company. He was active in outdoor sports. It appeared that he might be bisexual, but no homosexual exposures were admitted. However, he had been very friendly with a fifteen-year-old boy, Carl, during the past two years, getting jobs for him, buying his clothes and feeding him. Bob admitted sleeping with Carl on several occasions, but they both denied sexual contact. Nevertheless, Carl had a positive blood test for syphilis in June, 1956, when operated on at a local hospital, and a generalized eruption of secondary syphilis on August 16. We therefore assumed that exposures had taken place.

Bob went to a private physician because of a rash on the buttocks and genitals and anogenital sores. A diagnosis of late secondary syphilis was established on August 16. The blood test was positive. Bob admitted 3 female contacts:

6c-i. Helen (contact 6). Exposure in January.

6c-ii. Tina was twenty years old and single. Bob picked her up on the street had an exposure with her in his car on several occasions in May. She was not infected.

6c-iii. Ursula was a nineteen-year-old girl whom Bob met in a restaurant in another state in March. Exposure occurred in an automobile. This contact could not be located.

6c-iv. Carl (above).

7.* Irma was an attractive, well built, thirty-nine-year-old widow with nine children, three of them illegitimate. Two of these children were married, four were in school, and the three illegitimate ones were of preschool age. Irma was on general relief and lived in what was nearly a slum area. She was well known to all the "sporting" men in the area, and they visited her at all hours of the day and night. No one was turned away. Irma was not a prostitute and would not take money for her services. She had exposures with Boris from October 1 to 6. On January 10 she was found to be infected with early latent syphilis, as evidenced by repeated positive blood tests for syphilis and negative spinal-fluid Wassermann and negative physical examination. Treatment was begun on this date. She admitted 8 contacts besides Boris:

7a. Quentin, Irma's son-in-law, was twenty-eight years old and married to Lorraine (contact 7a-ii). He was immature, thin and undernourished, and was sullen and taciturn. When he came to the clinic on January 13 he had a profuse urethral discharge, and a diagnosis of gonorrhea was established on January 13. Blood tests were

On the basis of the information available, the most likely source of the syphilis outbreak was Irma, the thirty-nine-year-old widow. To Karen goes the dubious honor of initiating the epidemic of gonorrhea.

positive, and on March 13 a diagnosis of early latent syphilis was made. However, Quentin denied any sexual exposures, stating that his wife was pregnant and he did not believe in married people "running around." He had had "nothing to do with his wife for over two months." On his third visit he broke down and admitted that he had had an exposure some time in December with Karen in his car. While receiving treatment he told the nurse epidemiologist that he had been doing a lot of thinking and had decided to tell who his other contact was. He then named his mother-in-law (Irma). This exposure took place between Christmas time and New Year's at Irma's home. Later when Quentin's wife's blood was found to be positive for syphilis, he admitted relations with her in late February.

7a-i. *Karen, a nineteen-year-old single woman, was examined at the clinic on December 13 and had positive smears and cultures for gonococci. She was treated for gonorrhea on December 20. Repeated blood tests for syphilis were negative. She named 5 contacts:

7a-i (1). Peter (contact 7b) had an exposure with Karen on November 20, and a urethral discharge developed four days later.

7a-i (2). Walter, twenty-four years old and single, had an exposure with Karen on December 8. Two days later a urethral discharge developed and on smears and cultures proved to be due to gonorrhea. He was treated on December 10.

7a-i (3). Xavier, twenty-two years old and single, had an exposure with Karen on December 6, and a urethral discharge developed three days later. He was treated for gonorrhea on December 9.

7a-i (4). Sam, twenty-four years old and single, had an exposure with Karen in November. He was not infected with either gonorrhea or syphilis.

7a-i (5). Quentin (contact 7a) had an exposure with Karen early in December.

7a-ii. Lorraine, the seventeen-year-old wife of Quentin and the daughter of Irma, was found to have gonorrhea on January 13 and secondary syphilis on April 7. A son was born on January 3; repeated examinations of the infant were negative for syphilis. When Lorraine learned that her husband had had relations with her mother she told him she did not like it, but after all, she could not blame him too much — "that's nature."

7b. Peter was twenty-four years old, unmarried and good-looking. He frequented the barrooms and was often arrested for fighting when drunk. He was extremely co-operative and was directly responsible for getting to the clinic some of the contacts mentioned by Irma. Peter admitted exposures with Irma during September and October. He had gonorrhea on November 24 and was treated by a private physician. He had a second attack on February 8. Repeated blood tests for syphilis were positive, and the spinal-fluid serologic reaction was negative. On February 14 he was diagnosed as having early latent syphilis and treatment was begun. Peter gave the names of 2 contacts besides Irma:

7b-i. Janet (contact 6b-i). Exposure in January.

7b-ii. Karen (contact 7a-i). Exposure on November 20.

7c. Oscar (contact 6b) had repeated exposures with Irma in December.

7d. Robert was twenty-six years old and the father of one of Irma's three illegitimate children. This child was born in August, 1954, when Irma's prenatal blood test was negative. Robert continued exposures with her up to October, 1955. Repeated examinations were negative.

7e. Sam, twenty-two years old and single, had contacts with Irma from September to October. He was not infected with either gonorrhea or syphilis.

7f. Tom, nineteen years old and single, had contacts with Irma in July. He was not infected.

7g. Ulysses, eighteen years old and single, also had contacts with Irma in July. He was not infected.

7h. Victor, twenty years old and single, had contacts with Irma in October. He was not infected.

• • •

On the Role
of Sex in Nature

Sex is common. Virtually all animals reproduce sexually, although some, like aphids and mites, have periods of asexual reproduction and a few organisms have even foregone sexual reproduction entirely. There are species of flies, for example, in which males are both unknown and unnecessary. Females of certain small fish, to cite a second example, allow (invite ?) themselves to be fertilized by males of other species as a stimulus to reproduction but do not use the sperm of these males. Plants reproduce sexually. Each flower generally has its male and female parts, although in an occasional species, like the date palm or the Christmas holly, separate individuals produce male and female flowers. Plants, of course, commonly retain very efficient means for reproducing asexually by tubers, bulbs, runners, cuttings, and the like. The more we learn of microscopic organisms, the more we realize that these, too, have means for reproducing sexually. Similarly, in the viruses themselves we encounter sex on the molecular level.

Quite obviously, both lower and higher organisms can utilize sexual reproduction or leave it alone. Why have so many different forms opted in its favor even at the expense of losing alternative, asexual means of reproduction? This is the problem for this essay. First, however, a clear statement must be made about "sexual reproduction" and "sex"; these words do not have the same connotations for all persons.

The geneticist, when operating professionally, regards "sex" as a mechanism whereby one individual is enabled to utilize genetic material from two different sources — generally his parents, although "parents" does not neatly fit the sexual process of microorganisms. All of the elaborate trappings that signify sex to the nongeneticist are only means by which the DNA of two individuals is brought together in a single cell. How remarkable! The mechanisms — calls, songs, colors, odors — by which individuals are held together in clusters or are brought together at the right moment in pairs, the elaborate costumes that identify the two sexes, the ritualistic courtships of many animals (including man), the mating procedures themselves — all of this is a superstructure that has been erected in order that DNA molecules may generate new combinations from old ones.

The extent to which sexual reproduction funnels genetic information to a particular individual can be appreciated by looking back through time at one's ancestral tree. Using myself as an example, my father was born in New York City and my mother in what was then the Territory of South Dakota; my four grandparents were from rural Pennsylvania and upstate New York; and my eight great-grandparents were scattered between two continents. My children would trace an even more geographically divergent pedigree since their mother's family is Central European in origin. The point is that an individual born of sexual reproduction looks back on a branching and rebranching pedigree that, as it leads back in time, spreads out to involve more and more of the species. The genetic material of these various sources, looking now from the past toward the present, has come together and recombined to produce the genetic endowment of the individual under discussion.

A bacterium that has arisen following a long regime of asexual divisions does not have a pedigree such as that described in the preceding paragraph. In each preceding generation, a bacterium has had but *one* ancestor if reproduction has been truly asexual. Two individuals cannot possibly contribute jointly to the genetic material of a subsequent one in an asexual organism. If we look from the past towards the present, we shall see that some bacteria have left many descendants whereas others have left none. In the opposite direction, however, there is but one ancestor in each generation — an unbroken line of single individuals extending far back into time.

The possibility opened to a species by sexual reproduction and missing in its absence is genetic recombination. Asexual organisms put together valuable collections of genes the hard way — one gene at a time. Each new mutant form, in order to spread or even to be retained in the population, must improve on what is already present. Sexual reproduction permits genes of various sources to come together and form immense numbers of combinations; some of these may be useful and will tend to increase in frequency. Although it is true that subsequent recombination will tend to destroy the very same combinations, it is also true that the valuable ones will become more and more frequent, and easier and easier to assemble from miscellaneous fragments. Evolution is a much faster process in a sexual than in an asexual species.

A number of forms alternate between sexual and asexual reproduction; wheat rust is one of these. A variety of rust capable of infecting the wheat in the Midwest spreads through thousands of acres of wheat land by asexual reproduction. Wheat production drops tremendously as a consequence. The farmers retaliate by sowing only resistant wheat on which this variety of rust is unable to grow. Wheat production rises again. On barberry bushes, however, the rust undergoes sexual reproduction, the reproduction that generates new varieties. Should one of these new varieties of rust be capable of infecting the rust-resistant wheat, the infection spreads throughout the wheat-growing states again by asexual reproduction. Asexual reproduction is an excellent way of producing large numbers of identical individuals and, consequently, is an excellent procedure for exploiting an environment of plenty. Sexual reproduction produces new genetic combinations; it is best for solving hitherto unsolved problems posed by a deteriorating or hostile environment.

Most higher organisms reproduce sexually. All of them have a life cycle that includes a period in which two complete sets of genes, maternal and paternal, are present. The bulk of things we can see — trees, smaller plants, people, and animals of all sorts — represent a stage of life in which the individual's cells have two sets of genes. The stage in which cells carry only one set is reduced to the germ cell stage in animals (sperm of males, eggs of females) or to the very short pollen and egg generations of higher plants. In contrast with higher organisms, lower ones like bacteria and fungi emphasize the stage in which only a single set of genes is found. Fertilization of an "egg" by a "sperm" in the case of a fungus produces a cell that contains two sets of genes. Recombination occurs within that cell, and the next two cell divisions immediately reduce the number of sets of genes per cell to one, the typical number for the growth period that follows. Bacteria possess a number of tricks by which one individual can come to possess genes from another and to incorporate these into novel combinations. Two bacteria can fuse (a true sexual process), and the DNA of one can migrate into the other; DNA can be released into the surrounding medium by a disintegrating cell and penetrate another cell more or less as a collection of "naked" genes; or DNA can be carried inadvertently from one cell to another by infectious virus particles.

The widespread occurrence of sexual reproduction attests to the evolutionary importance of gene recombination. Certain plants have forgone cross-fertilization by adopting self-fertilization; sex is retained, but it no longer serves its genetic role in these cases. Self-fertilization in most of these instances is obviously a characteristic that has arisen only recently because the complicated machinery for cross-fertilization is still present, but it is not working. For example, a self-fertilizing plant may possess colorful flowers that attract bees and other natural pollinators even though fertilization is completed before the flower opens.

The elaborate machinery that the nongeneticist associates with sex also attests to the importance of sexual reproduction. A great deal of all communication in the natural world, a great many of all aspects of behavior, and a considerable amount of the complex organization of communities is directly concerned with bringing together DNA of diverse origins. This edifice of community organization and reproductive behavior did not arise spontaneously; on the contrary, it has been assembled gradually by selective processes rooted in reproductive success.

Cousin Marriages:
An Essay in Three Parts

1. Mutant genes in populations

Genetic disorders pose a serious public health problem. The problem lies not in the extinction of mankind but in the substantial frequency of genetically handicapped persons. Various panels of experts have cited 1 per cent to 5 per cent as the frequency of genetically caused abnormal births in human populations. This frequency — whatever its precise value — is constant because it is an *equilibrium* frequency; until something untoward happens, the population is in a steady state. One such untoward event would be an increase in the average radiation exposure of man; this exposure would increase the rate of gene mutation and, in turn, necessitate a higher rate of elimination of mutant genes (by the birth of genetically abnormal children) in order to reestablish the steady state once more. I have commented earlier on this possibility.* A second untoward event would be the inadvertent but widespread use of a chemical mutagen for industrial or other purpose. A number of chemicals that have proven to be powerful mutagens in lower organisms (organisms for which adequate experimental data can be obtained most rapidly) are available in railroad-tank-car lots because they have wide application in industry. These epoxy compounds and still other widely used chemicals must be scrutinized with utmost care to guard against the induction of insidious changes in man's genetic material.

What are the links between mutation and the frequency of genetically handicapped individuals in a population? What is the relationship between these two events — one, a change in molecular structure; the other, the death or serious malformation of a newborn child? This is the subject of my brief essay.

The input and outgo of mutant genes in a population can, as a first approximation, be compared to the swimmers at a large municipal pool on a midsummer day. Suppose a new arrival comes to the pool every minute. Suppose, too, that only 1 of every 100 persons at the pool will leave in a one-minute interval. When the pool first opens in morning, the number of

*See the essay entitled "Radiation and Future Generations" in Vol. II.

customers increases steadily because new customers arrive at a rate greater than that at which old ones leave. As the number of persons at the pool approaches 100, however, the number leaving for home each minute approaches 1, the same value as that at which new customers arrive. And so, once the day is well under way, there should be about 100 persons at the poolside at any moment. This group of 100 persons has a constantly changing composition; it represents a *dynamic* (not a *static*) equilibrium. Should the rate of new arrivals increase after lunch, for instance, to 2 per minute, the number of persons found at the pool will increase once more until it becomes 200. When 200 patrons are present, 2 persons will leave each minute and so the new, increased rate of arrivals have been matched, and the number of persons at the poolside is again constant.

The analogy of the swimming pool is adequate only as a first approximation; it describes matters reasonably well for the origin of mutations and their loss from a bacterial culture in which each individual bacterium has but one set of genes. The normal genes in a bacterial culture mutate to an abnormal form at a constant rate. A certain proportion of the abnormal genes are lost each generation because, for example, the growth rate of abnormal bacteria is reduced. The proportion of mutant genes found in the culture is constant and, furthermore, depends upon the two rates described here: rate of origin by mutation and rate of loss by differential reproduction.

To illustrate by analogy the gain and loss of mutant genes in man and other higher organisms would demand an unacceptably artificial example. The reason lies in the possession by man and these other organisms of *two* representatives of each gene and the need for both of these to be mutant in order that the individual exhibit grossly abnormal characteristics.

Suppose that a mutant form of a gene arises unexpectedly (= new mutation) in every millionth sperm or egg. Suppose, too, that an individual who receives a mutant gene from both his father and his mother dies of a severe genetic disease before reproducing. How frequent will the mutant gene be in the population under equilibrium conditions? The frequency of the gene will become constant when one person of every million conceived dies of this disease because the dead individual removes from the population one mutant gene that had previously been carried by an egg and another that had been carried by a sperm. Now, the one individual per million (1/1,000,000) equals the product of the equilibrium frequency times itself because it represents the chance that a rare mutant-carrying egg will encounter a rare mutant-carrying sperm. By trial and error if necessary or by algebra (x^2 = 1/1,000,000 and, therefore, x = 1/1,000) we can find that the equilibrium frequency of the mutant gene in the population equals one per thousand gametes. One sperm among one thousand and one egg among one thousand carry the mutant gene.

Although one per thousand is not a large number, it is still one thousand times greater than one per million, the frequency of affected children or the mutation rate itself. Furthermore, since the entire set of genes carried by a person consists of tens of thousands of individual genes, each person carries within him, more or less hidden from view, dozens, let us say, of mutant genes that are capable of leading to severe abnormalities if they were to be present in double dose. Not all these abnormalities need be lethal, for we include such things as deaf-mutism, albinism, dwarfism, and other clearly abnormal but nonlethal traits in our inventory.

Organisms, like man, that receive genetic material from two parents accumulate mutant genes under the impact of recurrent mutation. The accumulation proceeds until the accidental "matching" of similar mutant genes at fertilization brings about their loss at a rate equal to the rate at which they arise by mutation. Hence, the formation of genetically abnormal individuals is directly related to the rate of mutation; the one must equal the other in an equilibrium population. Furthermore, in the process of equalizing input and outgo of mutant genes, a relatively large store of mutant genes accumulate (like bathers at a municipal swimming pool on a hot summer's day) within the genetic material possessed by the population.

2. Children of cousin marriages

Genetically abnormal individuals, individuals suffering from hereditary diseases, are a relatively small fraction of a population such as that of the United States; estimates of their overall frequency generally lie between 1 per cent and 5 per cent. Hundreds of separate genetic traits are included, of course, within the general category "hereditary disease." The actual frequency of any one abnormality may be as low as one birth per 10 thousand, or one per million, or even lower. These are the affected children who, because of their physical defects and lowered reproductive success, counterbalance the mutational origin of faulty genes.

Children whose parents are cousins are more likely to suffer from heritable diseases than are children born to unrelated couples. So much more so, in fact, that upon encountering someone afflicted with a rare genetic defect, it is a safe bet in most societies to *predict* that the individual's parents are indeed cousins. In this essay I want to explain why the children born of cousins are so much more likely to exhibit a genetic defect than are other children. I shall also report on some additional characteristics of seemingly normal children born of cousin marriages. I shall conclude the essay by mentioning that most plants and animals have devices by which self-fertilization and other close inbreeding is avoided.

A child born of cousins does not draw his two sets of genes from sources that are independent of one another. The reason is that his parents share two grandparents in common. Ordinarily, couples provide their children with eight great-grandparents, but married cousins provide only six.

The following diagram will help in tracing a mutant gene from one great-grandparent to the child whose parents are cousins:

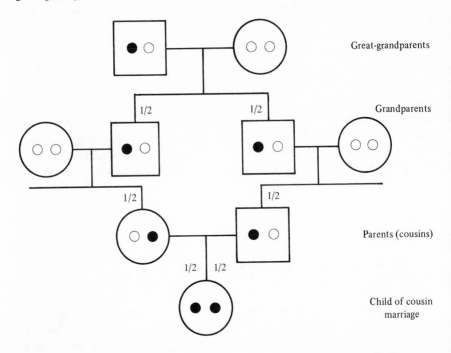

The solid circle represents a mutant gene carried by one of the great-grandparents; in each generation this gene has one chance in two of being passed from a carrier parent to his offspring. There is one chance in four that the two grandparents in the figure will receive mutant genes from the one great-grandparent. There is one chance in four that the two grandparents will pass the gene on to the two cousins who will eventually marry. There is also one chance in four that both cousins will pass the mutant gene to the same genetically defective baby. The cumulative probability of a defective child arising as shown in the diagram equals $\frac{1}{4} \times \frac{1}{4} \times \frac{1}{4} = \frac{1}{64}$.

If the mutant gene carried by the great-grandparent has a frequency of one per thousand in the population, two grandparents have very nearly four chances in a thousand of carrying one of these mutant genes because two persons possess four chromosome sets or four copies of each gene. The probability of obtaining a child exhibiting this particular mutant trait as an outcome of cousin marriage is, consequently, $4 \times 1/1,000 \times 1/64 = 1/16,000$. This is considerably greater than the one chance in a million that applies to children of unrelated parents.

I suggested earlier that in most societies we might wager that a person afflicted with a rare genetic disorder is the child of a cousin marriage. This is a rather safe bet because the probability that the mutant gene will be expressed in the child of a cousin marriage is greater than in a child born to unrelated parents. Not only is the first probability greater, but it becomes increasingly so in the case of rarer and rarer mutant genes. As long as one or more marriages of each hundred are cousin marriages, the defective children of these marriages will form a substantial proportion of all genetically defective children in a population.

Not all mutant genes cause gross defects to their carriers. Nevertheless, one-sixteenth of all genes carried by a child of a cousin marriage will be identical (the paternal and maternal copies of these genes will be identical) because those carried by both the mother and father have been handed down directly from a common great-grandparent.

Other than to increase the frequency of relatively rare defects, how does the close kinship of parents affect the health of the children they produce? There have been a number of studies that supply answers to this question; an especially fine one has been made on Japanese children by Drs. William J. Schull and James V. Neel.* Children born to parents who are cousins are impaired in many respects when compared to children born of unrelated parents. Early mortality is higher in children of cousin marriages; they possess a higher frequency of major and minor defects many of which are known to have genetic bases; the age at which they first walk or talk is delayed; physically they are a bit smaller; and their school grades are somewhat lower. These effects are so small that measurements on hundreds of children were necessary to demonstrate them conclusively. Nevertheless, they amount in most cases to a decrement of 1 per cent to 3 per cent. The early mortality for the children of cousin marriages is increased about 15 per cent above that of children born to unrelated parents.

In the world of wild species of plants and animals, many reproductive and behavioral devices exist that assure outcrossing instead of inbreeding. The pollen of a flower may ripen before the stigma of the same flower is receptive; the pollen of one flower, consequently, will function only on the stigma of another. The stigma may be placed so that bees and other insects deposit foreign pollen on it before they encounter the pollen of the same flower; again, the seeds will arise by outcrossing and not be inbreeding. The spider web of scent trails that radiate from a rural house and advertise to passing dogs that the resident female is in heat are all too effective in luring the male dogs of the neighborhood. I still recall the small beagle that first arrived at my doorstep, followed by terriers and mongrels, and finally by an enormous German shepherd who clawed all night at the window sills trying to find an entrance into my house.

Man, too, has devised rules regulating marriages and the choice of marriage partners. These rules have been devised for many different reasons of course. They control the flow of property and other wealth from generation to generation; they influence political power; and in the few remaining royal families they assure the often unemployed husband of an understanding and

*William J. Schull and James V. Neel, *The Effects of Inbreeding on Japanese Children* (New York: Harper and Row, 1965).

sympathetic wife. More often than not, however, the marriage customs also assure that husband and wife are not closely related. By means of shrewd observations, the wise men of nearly all societies have noted that close inbreeding often leads to the production of physically and mentally handicapped children. It is rare indeed to find marriages closer than that of first cousins condoned by society; within the Roman Catholic church, the marriage of first cousins requires a special dispensation.

3. Genetic counseling

The presence in human populations of mutant genes capable of killing or, sometimes worse, of crippling their carriers both mentally and physically casts a pall over many marriages. The increased chance that the child of a cousin marriage will carry a double dose of one of these genes has been recognized in many societies as an unacceptable risk and such marriages are banned. Cousins do fall in love and marry, nevertheless. The purpose of this brief essay is to describe services that are available for counseling those persons who have or, if marriage plans are yet to be carried through, may have genetic problems. These services are available at many of the medical schools associated with the larger private and state universities.* The American Eugenics Society exists to disseminate accurate genetic and demographic information rather than to promote eugenic measures as such; by educating the public at large in genetic matters it serves an extremely useful purpose.

Genetic counseling consists of much more than a recital of probabilities to couples who have already had one and who may have still more abnormal children. These bare figures can be obtained from nearly anyone who has taught — or, indeed, has taken — a reasonable genetics course. Counseling entails a thorough explanation of the nature of heredity and inheritance to the couple involved, an examination of the problem immediately at hand, and a careful and compassionate search for possible courses of action and their probable outcomes. The association of the counseling service with a medical school is especially important in the review of possible courses of action; these change so rapidly with modern medical advances that one must be a professional to keep abreast of them.

The goal of good counseling is to confer the greatest possible happiness on

---

*Addresses and other pertinent information are available from the American Eugenics Society, 230 Park Avenue, New York, N. Y. 10017.

those seeking advice and to do this in a way that assures the greatest benefit to the population and to future generations. I could, for example, advise all persons who carry hidden deleterious mutant genes not to marry others who carry the very same genes. This advice would preclude the birth of individuals carrying a double dose (one mutant gene from their mother and a second from their father) of the same mutant gene; it would be reasonably good advice if I were concerned exclusively with the present generation. As long as mutation occurs, however, elimination of mutant genes must also occur or the frequency of the mutant gene in the population will increase. My advice, therefore, is nothing but a stop-gap measure that reduces the number of genetically abnormal persons just for the moment.

What sort of advice would be better? In my estimation, advice that, if followed, would counterbalance mutation by removing mutant genes from the population would be much better, provided that such advice can be found. Some of the persons carrying abnormal genes may not be particularly interested in having children. I would urge these couples to make a conscious decision and take definitive steps to remain childless so that they would not have children thoughtlessly or through carelessness. In still other instances, the genotype of the unborn child can be determined at a very early stage in development; I would recommend therapeutic abortions of recognized carrier embryos as a means of eliminating the causative mutant genes.

In the past, childless couples have been the target of community gossip and have had to tolerate endless personal interrogation at the hands of local busybodies. This may change in the near future so that childlessness would become an acceptable eugenic procedure. I base this claim regarding childlessness on the grounds that no other problem confronting mankind can be solved satisfactorily until the problem of stabilizing population size is also solved. Zero growth rate is the only tolerable growth rate for an already oversized population; the number of persons in human populations must be stabilized. When this point sinks home, society will realize that each year during which a woman remains childless is a year of valuable service to the community. We shall eventually come to pay for this service, I am sure. Under this system, many couples would supplement their incomes by annual stipends gained in avoiding childbirth that year. With the added incentive that their children are likely to be defective, parents with genetic problems would probably join the ranks of (paid) childless couples.

Genetic counseling, in addition, serves an important role when it frees one side or the other of a family from the onus of "bad blood." The number of times one marriage partner wheels accusingly on the other to say that *that* trait comes from *your* family is simply amazing. Often the accusation itself is true, but the implication behind it never is. If the trait is one caused by a mutant gene, it should be obvious that the family did not deliberately choose to possess it. Furthermore, only chance has prevented equally undesirable traits on the opposite side from appearing in the children. We all carry our genetic defects. They were handed down to us from our parents by chance, and we shall pass them to our children in the same manner. No family is genetically perfect; no one is "to blame" for the genetic characteristics of a child.

Genetic counseling is valuable, too, in the case of *all* cousins who contemplate marriage. We have seen that there are penalties that befall children born of cousins, penalties that range from a greater chance of early death to lower school grades. Counseling in this case can establish whether children are important to the couple or whether adopted children would be acceptable. By examining the family pedigree the counselor can see what obvious genetic defects, if any, have appeared in the past. It may happen that the cousins involved can be tested specifically for the presence of genes responsible for certain of these defects. Finally, counseling can evaluate the emotional state of the couple and weigh it (1) against advice to forego marriage and (2) against the slight genetic harm that might befall their children in the apparent absence of gross defects in the family pedigree.

Genetic counseling, properly done, serves many purposes. It relieves tensions between estranged parents. It educates. It reassures those who are, or think they are, carriers of genetic disorders. It abolishes ill-defined fears and replaces them with concrete problems and alternative solutions that can be reviewed rationally, one by one. The replacement of worry by purposeful action may be the most valuable service rendered by the counselor. And here we encounter Sandburg saying, "I am sure that if I had been struck as deep as he [a childhood friend] was by the infamy of this one event so memorable to him, I would go on a search. I would seek all items of information that would throw light on what happened afterward. . . ." The well-trained counselor, using all of the genetic and medical knowledge at his disposal, joins with the troubled couple and aids in that search.

The Biology of Birth Control:
An Essay in Three Parts

1. Hormones: internal messengers

Life requires organization. A cell is not a tiny bag of fluid; it is a microcosm of compartments, membranes, organelles, and other structures. The collisions of molecules that are necessary for the occurrence of vital chemical reactions are largely the result of thermal motion, diffusion, and Brownian movement. Membranes separate compartments of the cell within which different, and perhaps antagonistic, reactions predominate; membranes, by "capturing" various molecules, can accelerate certain reactions that would proceed very slowly if the molecules were left to encounter one another by chance alone. Small as a single cell may be, for some molecular reactions it represents a universe of almost limitless dimensions.

The reactions that take place within a cell are controlled reactions. The cell can manufacture many chemical compounds. In making some substances it uses up others. These transformations are carried out in an orderly way through a variety of feedback circuits. In brief, each substance that is synthesized can, when it is present in sufficient supply, turn off machinery responsible for its own manufacture. The depletion of one substance during the synthesis of another (or its depletion by any other means) can reactivate the machinery required for its manufacture. The various substances within the cell maintain a continuous inventory of their own supply. Depletion stimulates synthesis or importation; an excess supply inhibits synthesis.

A large multicellular organism — animal or plant — needs regulatory machinery suitable for coordinating its entire body. The human body, for example, is not a haphazard collection of cells. It is a well-organized collection of cells, each of which is specialized to perform (usually in company with many others but sometimes alone) a particular function. Paradoxically, it is a collection of disparate cell types, all of which have arisen from preexisting, less diverse cells. The start of each individual's development is, of course, a single cell — the fertilized egg.

We understand in part the intricate mechanisms by means of which developmental processes are controlled; to describe what is known would require not an essay but a large book or perhaps a library of large books. The use the body makes of chemical messengers for the control of coordination of various organs can be described adequately in a few pages if the description is limited to broad outlines. I propose to provide this outline. From this general introduction we can work our way to a discussion of the physiology of birth control.

Within the body, two systems have ramifications that permeate every nook and cranny, every finger and toe; these are the nervous system and the circulatory system. The one is obviously a coordinating system much like a communications network but one that includes not only the wires and amplifiers but also the officials who run the network and who are empowered to receive messages, make decisions, and transmit orders. The second is a transport system; it moves oxygen, sugar, vitamins, and essential elements to all cells of the body and removes carbon dioxide and nitrogenous wastes. Because it does lead to every cell, however, the circulatory system can also be — and is — used as a communications system. The nervous system controls reactions that must be made in split seconds (we draw our hand back from a hot stove *before* we sense either heat or pain). The communications that travel in the bloodstream regulate processes that take minutes, days, months, or even years to react in full measure.

Communication between different parts of the body by way of the circulatory system is carried out by means of chemical compounds known as *hormones*. These substances, of diverse chemical structure, are synthesized in any one of several glands. At a certain time or in response to a particular signal one of these chemicals is dumped into the bloodstream and is carried to one or more target organs, which are then stimulated to respond in an appropriate manner.

Adrenaline is an excellent example of a hormone; its use is largely reserved for dire emergencies. Suppose you encounter something frightening that startles you. Two small glands near your kidneys inject about ½ gram of adrenaline into your bloodstream; this is done in response to a stimulus received by way of the nervous system. Almost at once the adrenaline is carried to your heart, which in response begins to pound furiously. Surface blood vessels are constricted, as are those serving the digestive system; blood is circulated almost exclusively through muscles and lungs. Bronchial tubes are dilated so that breathing is unimpeded. Cells in the liver are stimulated to produce an enzyme that converts stored glycogen into sugar that is available to the muscles for immediate use. The entire physiological machinery of the body that might be called upon for either combat or sustained flight is brought into readiness by adrenaline, a Paul Revere of chemical messengers.

The essentially instantaneous communication of the nervous system was contrasted earlier with the slower, often sustained, communication of hormone-producing glands. This comparison is not quite legitimate. For example, the adrenals, that is, the part that produces adrenaline, appear to arise from primitive nervous tissue. Furthermore, what appears to be the master regulator of the body, a part of the brain called the hypothalamus, and the large anterior part of its main target organ, the pituitary gland, are also derived from primitive nervous tissue. Nervous tissue is now under extensive investigation in an effort to discover roles that it might play other than that of a mere conductor of electrical impulses. It appears to possess secretory and transport properties not seriously considered in the recent past, a matter to be raised again in the section on behavior in this volume.

One hormone can lead to the production of another; a hormone can bring about an end to its own production. Hormones are the components of feedback systems. Some of these systems are dampened ones that maintain conditions on an even keel. Household examples are the "heating" thermostat that controls the furnace and the "cooling" thermostat that controls the airconditioner. The automatic pilot is also a device that relies on negative, or dampened, feedback. Comparable hormonal systems operate within the body. The hypothalamus stimulates the anterior pituitary to secrete a hormone that acts on the testes; under the influence of the anterior pituitary hormone, the testes secrete androgen, the sex hormone responsible for the secondary sexual characteristics of the mature male. Androgen also depresses the stimulating action of the hypothalamus. Hence, the body approaches an even keel under a system of negative feedback interactions.

Feedback mechanisms need not lead to immediate stability. They may be arranged so that the system under control moves more and more in a given direction. In man's technology, such positive feedback mechanisms are found in amplification systems, which are designed to magnify small events into large measurable or observable ones. Often, however, positive feedback reactions spell disaster. The erroneous attachment of a heating thermostat to an airconditioner or of a cooling thermostat to a furnace is an example. Under certain flight conditions, to cite a more serious instance, the engine pods of early Electra planes oscillated with increasing (rather than dampened) violence; on this account a number of these planes lost their wings in flight.

The ovulation of female rabbits represents a crescendo following a series of positive feedback interactions. The system begins at mating. Copulation stimulates the doe's nervous system, which in turn stimulates the pituitary. The pituitary secretes a hormone that stimulates the ovaries to secrete a second, pituitary-stimulating hormone. The secretions involved in the ovary-pituitary cycle increase in quantity after each go-around until ovulation occurs. In the absence of information about internal secretions, the casual observer would claim that ovulation in the rabbit is induced by copulation, and indeed it is. Copulation is only the first small stimulus in a subsequently self-amplifying system however.

Oversimplified interpretations, such as the notion that copulation leads to ovulation in rabbits, are not entirely wrong; they merely underestimate the true complexity of an existing situation. In truth, however, most of man's understanding of events is oversimplified. Gross causes of effects are recognized reasonably quickly but the more complex details are filled in only after carefully designed studies have been completed. I shall not seriously mislead anyone, therefore, if I use the kidney as still another illustration of a simple, negative feedback circuit. Normally blood enters the kidney by way of the renal artery;

the kidney removes various toxic substances, and the blood then leaves by way of the renal vein. Suppose the renal artery to each kidney is partially closed experimentally by means of a clamp so that the arterial blood arrives at the kidney with substantially reduced pressure. The kidney responds by secreting into the bloodstream a substance known as renin. Within the bloodstream renin helps produce a second substance known as angiotensin. The second substance, by constricting the blood vessels of the body, causes the blood pressure to rise. Because of the increased pressure, the normal supply to the kidney is restored.

The kidney, a vital organ with the responsibility for removing poisonous wastes from the bloodstream, can control its own blood supply by means of a hormone that it produces under the influence of reduced blood pressure. The details of this feedback circuit are more complicated than I have suggested; nevertheless, the circuit illustrates the self-regulating interactions that are responsible for the continued, coordinated operation of a living body and for the maintenance of life.

2. The female reproductive cycle

To many persons birth control means the absolute prevention of births, much as temperance for others means complete abstinence. The production of offspring, however, has been an essential aspect of life from the beginning of time. All other functions of existing organisms are, by definition, subordinate to this one; otherwise, the organisms would no longer exist. Hence, it should be no surprise to find that the young of all species are produced with as much care as the species can muster. In man, the normal reproductive process is under the control of a meshwork of hormonal actions and interactions. The control of the female reproductive cycle, with special emphasis on the monthly menstrual cycle, is the topic of this essay. My theme, however, deals not so much with reproduction as with the biology of regulation, not so much with specific organs and tissues as with the interlocking system of chemical controls that is, in fact, responsible for reproduction. The better one control system is understood in connection with one set of functions, the easier it is to anticipate difficulties while disentangling the details of a second system. The better we understand the system that controls birth, the easier birth control becomes.

The hypothalamus, a small region at the base of the brain, was mentioned in the preceding essay as the body's master regulator. It has a vital part in the regulation of the reproductive cycle. The nature of the stimulus that initially activates the hypothalamus is unknown; but, whatever it may be, at some time between the 10th and 13th years of a girl's life, the hypothalamus stimulates a nearby gland (the anterior pituitary) that is also at the base of the brain to produce two hormones known as FSH (FSH = follicle-stimulating hormone) and LH (LH = luteinizing hormone). These hormones are released into the blood-stream. The target organ for FSH is the ovary. Under the influence of FSH (and probably LH as well) the ovary matures and produces two substances of its own, estrogen and progesterone. Estrogen is largely responsible for the secondary sexual characteristics of women: development of the breasts, growth of pubic and axillary hair, and broadening of the pelvis.

The regular monthly menstrual cycle proceeds more or less as follows:

1.   The hypothalmus stimulates the pituitary to produce FSH and LH.

2.   FSH (traveling in the bloodstream) stimulates the maturation of the ovary. As a rule, a single egg matures faster than the remaining ones; it is surrounded by a sheath of cells, the follicle.

3.   The developing follicle produces estrogen, a hormone that stimulates the initial thickening of the uterine lining.

4.   When the follicle has matured and has released its egg by rupturing (an event that requires the second hormone, LH, of the anterior pituitary), estrogen production has reached its peak. At this level in the blood it *inhibits* the hypothalamus from stimulating FSH cells within the pituitary, but it *stimulates* the second part of the hypothalamus to increase its production of LH.

5.   LH, circulating by way of the blood stream, stimulates the ruptured follicle in the ovary to develop into a small nodule of cells known as the *corpus luteum* (Latin for "yellow body").

6.   The *corpus luteum* secretes a hormone, progesterone, that stimulates the final thickening of the uterine wall, anticipating the arrival of a fertilized egg. Although the details are obscure, it seems that progesterone *inhibits* the FSH-stimulating function of the hypothalamus; this inhibition would guarantee that the newly released egg is the only mature egg available for fertilization at this time.

7.   In the absence of fertilization, the continued high level of proges-terone also leads to an inhibition of the LH-stimulating region of the hypothalamus; Since LH appears to be responsible for the maintenance of the *corpus luteum*, its removal would cause this body to shrivel and disappear together with its secretion, progesterone. (The *corpus luteum* in a sense turns

itself off much like those nonsensical black boxes, on sale at novelty stores, from which a small white plastic hand appears and turns off the box's control switch each time the switch is turned on.) The lining of the uterus is partially resorbed and partially sloughed off (menstrual bleeding).

8.   In the absence of either estrogen or progesterone, the FSH-stimulating region of the hypothalamus is freed of its inhibition so that it once more stimulates the pituitary to make FSH — an act that returns us to the first item of this list. In a woman the sequence of these events requires about 28 days to run their course; this is the monthly menstrual cycle.

This essay has been exceptional in the number of technical terms used. I apologize, but I have seen no alternative. Interlocking circuits are difficult to describe verbally even if they are simple electrical circuits. The notion of feedback appears deceptively simple until we spell it out in words. Nevertheless, remember that the system described above has only five parts:

Brain   1. Hypothalamus
                                         2. Pituitary (anterior part)
Ovary   3. Follicle (with egg)
                                         4. *Corpus luteum*
Uterus   5. Uterine lining

and four hormones:

                                         1. FSH (follicle-stimulating hormone)
                                         2. LH (luteinizing hormone)
                                         3. Estrogen
                                         4. Progesterone

If one remembers these nine items, it is not difficult to fit the parts and the hormones together on paper so that they carry out their monthly cycle.

There is a fascination about hormones. They carry out their tasks at fantastically low concentrations. In some manner just a few molecules bring about effects completely out of proportion to their actual quantities. How does this happen? What amplification system is used? In at least one case a hormone is known to activate an enzyme directly. In another case (adrenaline) the hormone stimulates the production of a cofactor for an enzyme that is already present but in an inactive form; the cofactor attaches to the enzyme and activates it. Thus, it appears that the large effect produced by a few hormone molecules — the amplification of the hormone's effect — is brought about by their mediating influence on enzyme systems. At times the enzyme systems themselves may also occupy strategic positions in the feedback-controlled, metabolic pathways of the cell. In this case, a few hormone molecules may shunt the major activity of the cell from one pathway into a second, prearranged but at-the-moment unused, pathway.

3. Artificial birth control

"Blessed are the women that are irregular, for their daughters shall inherit the earth." Those are the words by which Dr. Garrett Hardin has answered the proposition that the rhythm (or "safe period") method of birth control might serve to control man's numbers. The rhythm method is inadequate for controlling the growth of human populations for the same reason that DDT has proven to be inadequate for controlling insect pests and antibiotics for eradicating disease organisms: natural selection will negate its effects. A large proportion of persons born in a population practicing birth control by means of the rhythm method would be born by accident to women whose reproductive cycles were irregular and whose fertile periods could not be foretold accurately. To whatever extent such irregularity has a genetic basis, the population would come to contain more and more irregular women. The monthly female reproductive cycle under these circumstances could become a historic curiosity.

The rhythm method has not been a particularly successful procedure for spacing pregnancies. In fairness, however, we must admit that Dr. Hardin's inspired phraseology is equally applicable to all types of birth control methods. Blessed are the impetuous, . . . or the forgetful, . . . or the careless. . . . Each variation recognizes that under a regime of widespread birth control many children will be conceived by accident no matter what contraceptive procedure is used; to an extent that would depend upon the genetic basis of the underlying blunder, the population would evolve so that the frequency of errors would increase. Of all such errors, however, irregularity of the female reproductive cycle would appear to be the most directly dependent upon the individual's genetic endowment.

Artificial birth control consists of voluntary procedures that prevent the encounter of eggs and sperm, make such encounters uneventful, or destroy the fertilized egg before or after (very early abortion) implantation in the uterine wall. For centuries numerous devices and procedures have been used for these purposes; birth control has been in the past and is today practiced on a worldwide basis. It is not my purpose here to review contraceptive measures simply for the sake of listing them. I assume that the reader knows of both spermicidal jellies and mechanical devices, such as condoms, diaphragms, intrauterine devices (IUD's) and cervical caps, that are readily available. Instead, I shall describe the biological basis of some of the newer oral contraceptives (the "pill" for example) and relate these contraceptive procedures to the normal functioning of the reproductive cycle. My essay could as well have been entitled *"Hormonal aspects of birth control."*

The details of the monthly female cycle were outlined in the preceding essay. The hypothalamus stimulates the anterior pituitary gland to secrete FSH (follicle-stimulating hormone) and LH (luteinizing hormone). FSH stimulates the maturation of an egg in the ovary whereupon the follicle within which it develops secretes estrogen, another hormone. The estrogen causes the hypo-thalamus to stimulate the anterior pituitary to increase its secretion of LH to levels that cause both the rupture of the follicle (thus releasing the egg it contains) and the transformation of the ruptured follicle into a small cellular body known as the *corpus luteum*. The *corpus luteum* in turn secretes progesterone, still another hormone. Progesterone stimulates the thickening of the uterine wall in anticipation of the arrival of a fertilized egg; in the absence of fertilization, however, it inhibits further production of LH by the pituitary gland. As the supply of LH diminishes, the *corpus luteum* itself atrophies so there is no more progesterone (nor even the small amount of estrogen that the *corpus luteum* also secretes). The absence of both progesterone and estrogen causes the hypothalamus to stimulate the anterior pituitary to secrete FSH and the cycle begins anew. The entire cycle requires about 28 days.

The most commonly used oral contraceptives (such as Enovid and Norlutin) are synthetic progesterone-like chemicals (progestins) that inhibit the production of FSH by the pituitary gland; sometimes a mixture of estrogen and progestins is used for the same purpose. The inhibition of the hormonal secretions of the anterior pituitary inhibits in turn the maturation of an egg and follicle in the ovary and leaves the uterine lining rather firm and compact. The drug is administered for twenty days, during which the ovary remains quiescent. When the drug is withdrawn, the uterine lining breaks down; menstrual bleeding occurs, and then a new cycle is started by renewing the drug once more.

The effectiveness of this contraceptive procedure lies in the suppression of egg maturation. There is no mature follicle to rupture and no mature egg to be released each month. The effectiveness of the procedure is nearly 100 per cent. On the other hand, the upset in the normal hormonal cycle has a tendency to cause certain adverse side effects — some temporary and unimportant, others severe enough to preclude the continued use of these contraceptive chemicals. The most severe of the unwanted side effects of the estrogen-containing pills is in the formation of abnormal blood clots, *thromboses*. These clots form within the blood vessels; they may circulate within the bloodstream and block the small arteries leading to the lungs or to the heart muscle. Thromboses are often fatal. In fairness, I must point out that those women who develop thromboses after taking the pill are the same ones who risk having thromboses during pregnancy. This particular hazard, consequently, cannot be said to arise wholly with the pill.

Enovid and Norlutin interfere with the normal female reproductive cycle by inhibiting the release by the anterior pituitary gland of FSH and LH, two ovary-stimulating hormones. Ordinarily, this gland would release its hormones into the bloodstream, which would then carry them to the target organ, the ovary. The bloodstream is, as we have learned, the distribution system for all hormones. That hormones are free in the blood for a short time makes them susceptible, in theory at least, to direct chemical inactivation. A plant of our

Western prairies, for instance, yields a hormonal-inactivating agent of just this sort; this agent, taken orally, appears to inactivate pituitary hormones in the bloodstream. The side effects of a continued treatment of this sort are unknown. A hope has been expressed that undesirable effects will be minimized because this chemical does not interact with glandular tissue itself. On the other hand, its use would seem to call for the unceasing function of both the hypothalamus and the anterior pituitary because of the absence of estrogen and progesterone in the bloodstream.

The human egg is fertilized not in the uterus but in one of the two Fallopian tubes that lead from funnel-shaped openings in the body cavity near the ovary to the uterus. At the time the follicle ruptures, I might point out, the egg is discharged directly into the woman's body cavity. It enters the uterus only after being caught up by ciliary action on the surface of the large funnel-like end of one of the Fallopian tubes and by traveling down the short tube for the next two or three days. Fertilization occurs within the tube. By the time the egg enters the uterus it has divided several times and actually consists of a sizable number of cells.

A chemical has been discovered that in experimental tests with rats causes the destruction of the egg during the very early stages immediately following fertilization. In man a chemical of this sort would mean that a pill would be taken a day or two *after* intercourse rather than for twenty days of every month. Instead of establishing an artificial monthly cycle during which the egg is prevented from maturing, the "after-the-fact" pill would simply guarantee that the fertilized egg, if one existed, would be promptly destroyed.

The final contraceptive chemical for our inventory is one that interferes with the implantation of the fertilized egg (by now a small cluster of cells) on the uterine wall. The uterus is prepared for acceptance of the egg by means of progesterone, the hormone produced by the *corpus luteum*. This hormone, besides causing the uterine wall to thicken, also induces the cells of the lining to form vacuoles filled with secretions of one sort and another. Contraceptive pills that inhibit the secretion of FSH and LH also inhibit the final stages of thickening of the uterine wall. These pills, therefore, would impede the implantation of any egg that might be released from the ovary by accident and subsequently fertilized. Other chemicals are known, however, whose effect is limited to an interference of the action of progesterone on the uterine wall. These chemicals keep the uterus in a state in which acceptance of the fertilized egg is unlikely or impossible. (IUD's, intrauterine devices, also prevent the firm implantation of the fertilized egg in the uterine wall. These devices work exceedingly well for most women although they have caused both discomfort and physical harm in others.)

Artificial birth control is becoming, paradoxically, a way of life. All contraceptive procedures except abstinence and the rhythm method are condemned by the Roman Catholic Church; the reasons for this opposition are set forth in the encyclical *Humane Vitae* of Pope Paul VI.* To a great many persons both within and outside the Church, the logic of neither the encyclical nor earlier birth-control bans seems compelling. The pill appears to be firmly entrenched in Western society as one of the many facets of modern life.

An essay on birth control is not necessarily the place for an editorial on the need to control births in a highly industrialized society. This need has been discussed in an essay entitled *"On population control"* as well as in other essays of Volume I. What I should do here is address some simple remarks to the young (and perhaps unmarried) reader but without appearing to sermonize. First, I want to reemphasize that epidemics, because they reflect the population growth of disease organisms, spread at exponential rates. The present epidemic of syphilis and gonorrhea has by now reached alarming proportions. This knowledge should destroy the peace of mind that must be an essential part of the pleasure that comes from any casual affair.

Second, the hormonal system that regulates the female reproductive cycle, like all of the body's regulatory systems, may switch into new and unexpected courses if it is too violently disturbed. Hormonal systems are interlocking dynamic systems that hold the body in a steady-state equilibrium. Despite the unprecedented amount of research that has been performed in developing the pill, it is still a biological sledgehammer. Perpetual pregnancy, the state induced by the progesterone-estrogen or progestin-estrogen combinations, has never been tested on a large scale during man's evolutionary history; that test is being made right now during the present decade. A young girl – especially an unmarried teen-age girl – who decides that contraception is a necessary feature of her life should think seriously (or, better, obtain professional advice) before embarking on a protracted course of hormonal control that eventually may extend over several decades. Mechanical devices (even the IUD) may easily suit her particular needs. As a present for the married girl who has had her family (two children if she heeds the call for zero population growth), her husband can obtain a vasectomy – a simple operation that frees the wife of any need to tamper further with hormonal systems that have been established by natural selection over hundreds of thousands of years.

*See Vol. I.

SECTION THREE

Communication

Introduction

"Communications broke down." "We failed to maintain lines of communication." These and similar excuses are offered when trouble explodes between couples, between generations, between races, or between nations. Whenever the explanation is correct, there exists cause for sorrow because a little additional effort could have prevented strife. Not all disputes arise from lack of communication however. Ours is not a peaceable kingdom, and so at times the aspirations of two opposing parties are incompatible. In that case, if negotiations fail and compromise proves to be impossible, conflict is inevitable. By maintaining communication at all times, however, the issues at stake become clearly defined so that the basis of the impending conflict is clearly understood by all concerned. Clarity of purpose offers the only means for evaluating expected gains in terms of the cost of conflict.

This section is built around a medley of topics related to communication in the animal kingdom. The selections that have been chosen to introduce the essays deal with a hierarchy of communication problems in human societies. The first, from James A. Michener's *Hawaii*, deals with a quaint courtship custom in an isolated Japanese village, a custom that in its exotic simplicity recalls the stereotyped mating dances of birds in a Walt Disney nature film. The pursuit of the young man by the girl's parents and their aroused neighbors could serve as a textbook example of ritualistic behavior – behavior that formerly served one purpose but, through gradual change over many years, has come to serve an entirely different one.

The second selection, from Alberto Moravia's *The Empty Canvas*, is a beautifully articulate account of what is generally described as "pulling teeth" – extracting information from the reticent. Communication between the couple in *The Empty Canvas* is, from the viewpoint of a would-be censor, hopelessly entangled with sex; I have chosen, however, not to assume that role. Among lower animals, of course, the bulk of communication centers on reproduction and the perpetuation of the species. In the case of man, the bonds between courting couples and among family members are established – Moravia's Cecilia notwithstanding – through communication.

Superimposed on the low hum of private communication among men, however, is the pandemonium of talk that cements the bonds of the community, the city, the state, the nation, and, indeed, the world. The third selection, chosen from William H. Whyte's *The Organization Man*, deals with the lowest of these societal bonds — the web of friendship in suburbia.

The essays that follow the three selections deal with what might be called the obvious aspects of simple communication. I have included other essays as well because this section is an appropriate place in which to discuss communication that is designed to mislead or, in the case of political activists, to rend society. Rhetoric, properly interpreted, is a form of communication. The final essay, for those who remain awake and alert, touches on the well-known sleep-inducing capacity of most technical reports and textbooks.

From the Inland Sea*

James A. Michener

In the year 1902, when the reconstruction of Honolulu's Chinatown was completed, one of the isolated farm villages of Hiroshima-ken, at the southern end of Japan's main island, stubbornly maintained an ancient courtship custom which everyone knew to be ridiculous but which, perhaps for that very reason, produced good results.

When some lusty youth spotted a marriageable girl he did not speak directly to her, nor did he invite any of his friends to do so. Instead, he artfully contrived to present himself before this girl a dozen times a week. She might be coming home from the Shinto shrine under the cryptomeria trees, and suddenly he would appear, silent, moody, tense, like a man who has just seen a ghost. Or, when she returned from the store with a fish, she would unexpectedly see this agitated yet controlled young man staring at her.

His part of this strange game required that he never speak, that he share his secret with nobody. Her rules were that not once, by even so much as a flicker of an eye, must she indicate that she knew what he was doing. He loomed silently before her, and she passed uncomprehendingly on. Yet obviously, if she was a prudent girl, she had to find some way to encourage his courtship so that ultimately he might send his parents to the matchmakers, who would launch formal conversations with her parents; for a girl in this village could never tell which of the gloomy, intense young men might develop into a serious suitor; so in some mysterious manner wholly understood by nobody she indicated, without seeing him or without ever having spoken to him, that she was ready.

Apart from certain species of the bird kingdom, where courtship was conducted with much the same ritual, this sexual parading was one of the strangest on earth, but in this village of Hiroshima-ken it worked, because it involved one additional step of which I have not yet spoken, and it was this next step that young Sakagawa Kamejiro found himself engaged in.

In 1902 he was twenty years old, a rugged, barrel-chested, bowlegged little bulldog of a man with dark, unblemished skin and jet-black hair. He had powerful arms which hung out from his body, as if their musculature was too great to be compressed, and he gave the appearance of a five-foot, one-inch accumulation of raw power, bursting with vital drives yet confused because he knew no specific target upon which to discharge them. In other words, Kamejiro was in love.

He had fallen in love on the very day that the Sakagawa family council had decided that he should be the one to go on the ship to Hawaii, where jobs in the sugar fields were plentiful. It was not the prospect of leaving home that had aroused his inchoate passions, for he knew that his parents, responsible for eight children and one old woman, could not find enough rice to feed the family. He had observed how infrequently fish got to the Sakagawa table — and meat not at all — so he was prepared to leave.

It happened late one afternoon when he stood in the tiny Sakagawa paddy field and looked out at the shimmering islands of the Inland Sea, and he understood in that brilliant moment, with the westering sun playing upon the most beautiful of all waters, that he might be leaving Hiroshima-ken forever. "I said I would go for only five years," he muttered stubbornly to himself, "but things can happen. I might never see these islands again. Maybe I won't plough this field. . . ever again." And a consuming sorrow possessed him, for of all the lands he could imagine, there could be no other on the face of the earth more exciting than these fields along the coastline of Hiroshima-ken.

Kamejiro was by no charitable interpretation of the word a poet. He was not even literate, nor had he ever looked at picture books. He had never talked much at home, and among the boys of the village he was known to be a stolid fighter rather than a talker. He had always ignored girls and, although he followed his father's advice on most things, had stubbornly refused to think of marriage. But now, as he stood in the faltering twilight and saw the land of his ancestors for the first time — in history and in passion and in love, as men occasionally perceive the land upon which they have been bred — he wanted brutishly to reach out his hand and halt the descending sun. He wanted to continue his spiritual embrace of the niggardly little field of which he was so much a part. "I may never come back!" he thought. "Look at the sun burning its way into the sea. You would think . . ." He did not put his thoughts into words, but stood in the paddy field, mud about ankles, entertaining tremendous surges of longing. How magnificent his land was!

It was in this mood that he started homeward, for in the Japanese custom all rice fields were gathered together while the houses to which they pertained clustered in small villages. Thus arable land was not wasted on housing, but the system did require farmers to walk substantial distances from their fields to their homes, and on this night little bulldog Sakagawa Kamejiro, his arms hanging out with their powerful muscles, walked home. Had he met some man who had earlier insulted him, as often happened in village life, he would surely have thrashed him then and there, for he thought that he wanted to fight; but as he walked he happened to see, at the edge of the village, the girl Yoko, and although he had seen her often before, it was not until then — when she walked with a slight wind at her dress and with a white working-woman's towel about her head — that he realized how much like the spirit of the land she was, and he experienced an almost uncontrollable desire to pull her off the footpath and into the rice field and have it over with on the spot.

Instead, he stood dumb as she approached. His eyes followed her and his big arms quivered, and as she passed she knew that this Kamejiro who was earmarked for Hawaii would watch her constantly throughout the following days, and she began to look for him at strange locations, and he would be there, stolid, staring, his arms hanging awkwardly down. Without ever acknowledging by a single motion of her own that she had even seen him, she conveyed the timeless message of the village: "It would not be unreasonable if you were to do so."

Therefore, on a soft spring night when the rice fields were beginning to turn delicate green, the sweet promise of food to come, Sakagawa Kamejiro secretly dressed in the traditional garb of the Hiroshima-ken night lover. He wore his best pair of trousers, his clean straw zori and a shirt that did not smell. The most conspicuous part of his costume, however, was a white cloth mask which wound about his head and covered his nose and mouth. Thus properly attired, he slipped out of the Sakagawa home, down a back path to Yoko's and waited several hours as her family closed up the day's business, blew out the lights and threw no more shadows on the shoji. When he was satisfied that Yoko had retired, with a reasonable chance that her parents might be sleeping, he crept toward the room which from long study he had spotted as hers, and in some mysterious way known only in the villages, she had anticipated that this was the night he would visit her, so the shoji had been left unlocked, and in a moment he slipped bemasked into the room.

Yoko saw him in the faint moonlight, but said nothing. Without removing his mask, for that was essential to the custom, he crept to her bed and placed his left hand upon her cheek. Then he took her right hand in his and held her fingers in a certain way, which from the beginning of Japan had meant, "I want to sleep with you," and of her own accord she changed the position of his fingers, which timelessly had signified, "You may."

With never a word spoken, with never a mask removed, Kamejiro silently slipped into bed with the intoxicating girl. She would not allow him to remove her clothing, for she knew that later she might have to do many things in a hurry, but that did not inconvenience Kamejiro, and in a few stolid, fumbling moments he made her ready to accept him. Not even at the height of their passion did Yoko utter a word, and when they collapsed mutually in blazing gratification and he fell asleep like an animal, she did not touch the mask, for it was there to protect her. At any moment in the love-making she could have pushed him away, and he would have had to go. The next day they could have met on the village street—as they would tomorrow—and neither would have been embarrassed, for so long as the mask was in place, Yoko did not know who was in her room. So long as the mask protected him, Kamejiro could not suffer personal humiliation or loss of face, for no matter what Yoko said or did, it could not embarrass him, for officially he was not there. It was a silly system, this Hiroshima courtship routine, but it worked.

When Kamejiro wakened, there was a moment when he could not recall where he was, and then he felt Yoko's body near his and this time they began to caress each other as proper lovers do, and the long night passed, but on the third sweet love-making, when the joy of possession completely captured them, they grew bolder and unwittingly made a good deal of noise, so that Yoko's father was awakened, and he shouted, "Who's in the house?"

And instantly Yoko was required to scream, "Oh, how horrible! A man is trying to get into my room!" And she continued to wail pitifully as lights flashed on throughout the village.

"Some beast is trying to rape Yoko-san!" an old woman screamed.

"We must kill him!" Yoko's father shouted, pulling on his pants.

"The family is forever disgraced!" Yoko's mother moaned, but since each of these phrases had been shouted into the night in precisely these intonations for many centuries, everyone knew exactly how to interpret them. But it was essential for the preservation of family dignity that the entire village combine to seek out the rapist, and now, led by Yoko's outraged father, the night procession formed.

"I saw a man running down this way!" the old woman bellowed.

"The ugly fiend!" another shouted. "Trying to rape a young girl!"

The villagers coursed this way and that, seeking the rapist, but prudently they avoided doing two things: they never took a census of the young men of the village, for by deduction that would have shown who was missing and would have indicated the rapist; nor did they look into the little barn where rice hay was kept, for they knew that the night fiend was certain to be hiding there, and it would be rather embarrassing if he were discovered, for then everyone would have to go through the motions of pretending to beat him.

In the hay barn, with chickens cackling, Kamejiro put on his pants, knocked the mud off his zori, and tucked away his white mask. When this was done, he had time to think: "She is sweeter than a breeze off the sea." But when he saw her later that day, coming from the fish stall, he looked past her and she ignored him, and this was a good thing, for as yet it was not agreed that Yoko would marry him, and if she elected not to do so, it was better if neither of them officially knew who had attempted to rape her. In fact, during that entire day and for some days thereafter Yoko was the acknowledged heroine of the village, for as one old woman pointed out: "I cannot remember a girl who screamed more loudly than Yoko-san while she was defending herself against that awful man . . . whoever he was." Yoko's father also came in for considerable praise in that he had dashed through every alley in the village, shouting at the top of his voice, "I'll kill him!" And farmers said approvingly to their wives, "It was lucky for whoever tried to get into that house that Yoko's father didn't catch him."

The Empty Canvas*

Alberto Moravia

Cecilia, as I think I have made clear, was not talkative, in fact it might be said that her natural inclination was to keep silence; but even when she spoke she contrived, as it were, to be silent at the same time, thanks to the disconcertingly brief, impersonal quality of her manner. Words, in her mouth, seemed to lose all real significance, and were reduced to abstract sounds as though they were words in a foreign language that I did not know. The lack of any kind of accent or dialect and of any inflection of social class, the complete absence of revealing commonplaces, the reduction of conversation to pure and simple declarations of incontrovertible facts such as 'It's hot today' — all these confirmed this impression of abstractness. I would ask her, for example, what she had done on the evening of the day before; and she would answer: 'I had dinner at home and then I went out with Mother and we went to the cinema together.' Now these words, as I immediately noticed — 'home', 'dinner', 'mother', 'cinema' — which in another mouth would have meant what they usually mean, and consequently, according to how they were uttered, would have made me see whether she was lying or telling the truth, these same words, in Cecilia's mouth, seemed to be nothing more than abstract sounds, behind which it was impossible to imagine the reality either of truth or of falsehood. How it was that Cecilia contrived to speak and at the same time give the impression of being silent, I have often wondered. And I came to the conclusion that she had only one means of expression, the sexual one, which however was obviously impossible to interpret even though original and powerful; and that with her mouth she said nothing, not even things concerned with sex, because her mouth was, so to speak, a false orifice, without depth or resonance, that did not communicate with anything inside her. So much so that often, looking at her as she lay beside me on the divan after our intercourse, flat on her back with her legs open, I could not help comparing the horizontal cleft of her mouth with the vertical cleft of her sex and remarking, with surprise, how much more expressive the latter was than the former — and with the same purely psychological quality as those features of the face by which a person's nature is revealed.

I had, furthermore, to discover what was concealed behind a remark such as: 'I had dinner at home and then I went out with Mother and we went to the cinema together'; whether, in fact, a dinner and a home, a mother and a cinema were really concealed behind the words, or possibly an appointment with the peroxide-haired actor; and thus, all at once, I was seized with a furious desire to know Cecilia better, for hitherto I had not taken the trouble to find out anything about her because, being under the illusion that I possessed her through our sexual relationship, I was under the illusion that I knew everything. For example, her family. Cecilia had told me, with her usual brevity, that she was an only daughter, that she lived with her father and mother and that they were not well off because her father was ill and had stopped working. I had been content with this information, almost grateful to her, in fact, for not telling me more, since the thing that mattered to me more than anything else was that she should come every day to the studio and make love with me. But from the moment when I had a suspicion that she was being unfaithful to me, and when this suspicion had suddenly transformed Cecilia from something unreal and boring into something real and desirable, I was filled with curiosity to know more about her home life, as though I hoped that a more thorough knowledge might enable me to achieve the full possession which sexual intercourse denied me. So I started questioning her, rather in the way in which I had questioned her about her relations with Balestrieri. Here, as an example, is one of our conversations.

'Your father is ill?'
'Yes.'
'What is he suffering from?'
'He's suffering from cancer.'
'What do the doctors say?'
'They say he's suffering from cancer.'
'No, what I mean is — do they think he can recover?'
'No, they say he can't recover.'
'Then he'll die soon?'
'Yes, they say he'll die soon.'
'Are you sorry?'
'Sorry for what?'
'That your father is dying.'
'Yes.'
'Is that all you can say?'
'What ought I to say?'
'But you're fond of your father?'
'Yes.'
'Well, let's go on to something else. Your mother — what's she like?'
'What d'you mean — what's she like?'
'Well, is she short, tall, pretty, ugly, dark, fair?'
'Oh well, I don't know; she's just like lots of other women.
'But tell me, what does she look like?'

'Goodness me, she doesn't look like anything.'

'Doesn't look like anything? Whatever do you mean?'

'I mean she doesn't look like anything in particular. She's just like anyone else.'

'Are you fond of your mother?'

'Yes.'

'More or less fond than of your father?'

'It's a different thing.'

'What does different mean?'

'Different means different.'

'But different in what way?'

'I don't know: different.'

'Well then, is your mother fond of your father?'

'I think so.'

'Why, aren't you sure?'

'They get on all right together, so I imagine they're fond of each other.'

'What does your father do all day?'

'Nothing.'

'What does nothing mean?'

'Nothing means nothing.'

'But people say "doing nothing" just as a manner of speaking, and then they really do all sorts of things, even if they're doing nothing. So your father doesn't work; what does he do instead?'

'He doesn't do anything.'

'That is to say — ?'

'Oh well, I don't know: at home he sits by the radio, in an armchair. Every day he takes a little walk — that's all.'

'I see. You live in a flat in the Prati district?'

'Yes.'

'How many rooms have you?'

'I don't know.'

'What d'you mean, you don't know?'

'I've never counted them.'

'But is it a big or a small flat?'

'So-so.'

'What does that mean?'

'Medium-sized.'

'Well then, describe it.'

'It's a flat just like lots of others; there's nothing to describe.'

'But I suppose this flat of yours isn't empty? There's some furniture in it?'

'Oh yes, there's the usual furniture, beds, armchairs, cupboards.'

'What sort of furniture?'

'Really I don't know; just like any other furniture.'

'Take the sitting-room, for example. You have a sitting-room?'

'Yes.'

'What furniture is there in it?'

'The usual furniture: chairs, small tables, armchairs, sofas — the same as in all sitting-rooms.'

'And in what style is this furniture?'

'Don't know.'

'What colour is it, then?'

'It hasn't any colour.'

'What d'you mean, it hasn't any colour?'

'I mean it hasn't any colour, it's gilt.'

'I see, but even gilt is a colour. D'you like your home?'

'I don't know whether I like it. In any case, I'm not there very much.'

I could go on, in this way, ad infinitum. But I think I have given a good example of what I have called Cecilia's abstractness. It may perhaps be thought, at this point, that Cecilia was stupid, and, anyhow, devoid of personality. But this was not so: the fact that I never heard her say stupid things was a proof, if nothing else, that she was not stupid; and as for personality, this, as I have already said, lay elsewhere than in her conversation, so that to report the latter without at the same time accompanying it with a description of her face and figure would be rather like reading an operatic libretto without music or a film script without the pictures on the screen. But I wished to transcribe an example of conversation mainly in order to convey the idea that Cecilia's way of speaking was thus formal and bloodless for the good reason that she herself was ignorant of the things about which I questioned her, just as much as I was and perhaps more so. In other words, she lived with her father and mother in a flat in Prati and had been Balestrieri's mistress; but she had never paused to look at the people and things in her life and therefore had never truly seen them, still less observed them. She was, in fact, a stranger to herself and to the world she lived in; just as much as those who knew neither her nor her world.

The Web of Friendship*

William H. Whyte, Jr.

In such characteristics as budgetism the organization man is so similar from suburb to suburb that it is easy to fall into the trap of seeing a "mass society." On the surface the new suburbia does look like a vast sea of homogeneity, but actually it is a congregation of small neighborly cells — and they make the national trends as much as they reflect them. The groups are temporary, in a sense, for the cast of characters is always shifting. Their patterns of behavior, however, have an extraordinary permanence, and these patterns have an influence on the individual quite as powerful as the traditional group, and in many respects more so.

Propinquity has always conditioned friendship and love and hate, and there is just more downright propinquity in suburbia than in most places. Yet in the power of the group that we see in suburbia we can see something of a shift in values as well. Just as important as the physical reasons is a responsiveness to the environment on the part of its members, and not only in degree but in character it seems to be growing.

In suburbia friendship has become almost predictable. Despite the fact that a person can pick and choose from a vast number of people to make friends with, such things as the placement of a stoop or the direction of a street often have more to do with determining who is friends with whom. When you look at the regularities of group behavior, it is very easy to overlook the influence of individual characteristics, but in suburbia, try as you may to bear this in mind, the repetition of certain patterns makes the group's influence abundantly obvious. Given a few physical clues about the area, you can come close to determining what could be called its flow of "social traffic," and once you have determined this, you may come up with an unsettlingly accurate diagnosis of who is in the gang and who isn't.

Now this may be conformity, but it is not unwitting conformity. The people know all about it. When I first started interviewing on this particular aspect of suburbia, I was at first hesitant; it is not very flattering to imply to somebody that they do what they do because of the environment rather than their own free will. I soon found out, however, that they not only knew quite well what I was interested in but were quite ready to talk about it. Give a suburban housewife a map of the area, and she is likely to show herself a very shrewd social analyst. After a few remarks about what a bunch of cows we all are, she will cheerfully explain how funny it is she doesn't pal around with the Clarks any more because she is using the new supermarket now and doesn't stop by Eleanor Clark's for coffee like she used to.

I believe this awareness is the significant phenomenon. In this chapter I am going to chart the basic mechanics of the gang's social life and what physical factors determine it, but it is the awareness of the suburbanites themselves of this that I want to underline. They know full well why they do as they do, and they think about it often. Behind this neighborliness they feel a sort of moral imperative, and yet they see the conflicts also. Although these conflicts may seem trivial to others, they come very close to the central dilemma of organization man.

For comprehending these conflicts, Park Forest is an excellent looking glass. Within it are the principal design features found separately in other suburbs; in its homes area the 60 x 125 plots are laid out in the curved superblocks typical of most new developments, and the garden duplexes of its rental area are perhaps the most intense development of court living to be found anywhere. Park Forest, in short, is like other suburbs, only more so.

Some might think it a synthetic atmosphere, and even Park Foresters, in a characteristically modern burst of civic pride, sometimes refer to their community as a "social laboratory." Yet I think there is justification for calling it a natural environment. While the architects happened on a design of great social utility, they were not trying to be social engineers — they just wanted a good basic design that would please people and make money for the developers — and some of the features they built into the units turned out to be functional in ways other than they expected.

But functional they have been. Perhaps not since the medieval town have there been neighborhood units so well adapted to the predilections and social needs of its people. In many ways, indeed, the courts are, physically, remarkably similar to the workers' housing of the fifteenth century. Like the Fugger houses still standing in Augsburg, the courts are essentially groups of houses two rooms deep, bound together by interior lines of communication, and the parking bay unifies the whole very much as did the water fountain of the Fugger houses.

Park Forest is revealing in another respect. There are enough physical differences within it to show what the constants are. When the architects designed the 105 courts and the homes area, they tried hard to introduce some variety, and because of differences in the number of apartments, the length of streets, and the way buildings are staggered around the parking bays, no two courts or superblocks are alike. Neither are they alike socially; some neighborhood units have been a conspicuous social success from the beginning, while others have not. There are more reasons than physical layout for this, of course, yet as you relate the differences in design of areas to the differences in the way people have behaved in them over a period of time, certain cause-and-effect relationships become apparent.

Let us start with the differences. In "spirit" they are considerable, and the lottery that takes place in the rental office when a couple is assigned to Court B 14 or Court K 3 is a turning point that is likely to affect them long after they have left Park Forest. Each court produces a different pattern of behavior, and whether newcomers become civic leaders or bridge fans or churchgoers will be determined to a large extent by the gang to which chance has now joined them.

Court residents talk about these differences a great deal. In some areas, they will tell you, feuding and cussedness are chronic. "I can't put my finger on it," says one resident, "but as long as I've been here this court has had an inferiority complex. We never seem to get together and have the weenie roasts and anniversary parties that they have in B 18 across the way." In other courts they will talk of their esprit de corps. "You would be lucky to get assigned here," says one housewife. "At the beginning we were maybe too neighborly — your friends knew more about your private life than you did yourself! It's not quite that active now. But it's still real friendly — even our dogs and cats are friendly with one another! The street behind us is nowhere near as friendly. They knock on doors over there." Community leaders explain that they have to become professionally expert at diagnosing the temper of different areas. In a fund-raising campaign they know in advance which areas will probably produce the most money per foot pound of energy expended on them, and which the least.

When I first heard residents explain how their court or block had a special spirit, I was inclined to take it all with a grain of salt. But I soon found there was objective proof that they were right. I began a routine plotting of the rate of turnover in each area, the location of parties, and such, and as I did so the maps revealed geographical concentrations that could not be attributed to chance. The location of civic leaders, for example. In the nucleus of courts that were settled first one would expect to find a somewhat higher proportion of leaders than elsewhere, but except for this I thought the distribution of leaders would be fairly random. Where potential civic leaders live, after all, had been determined more or less by chance, and three years of turnover had already shifted the personnel in every court.

But when I plotted the location of the leaders of the church civic organizations, certain courts displayed a heavy concentration while others showed none at all. The pattern, furthermore, was a persistent one. I got a list of the leaders for the same organizations as of two years previous and plotted their location. The same basic pattern emerged. Some of the similarity was due to the fact that several leaders were still hanging on, but there had been enough turnover to show that the clustering was closely related to the influence of the court.[1]

Other indexes show the same kind of contagion. A map of the roster of the active members of the United Protestant Church indicates that some areas habitually send a good quota of people to church while other areas send few. Voting records show heavy voter turnouts in some areas, apathy in others, and this pattern tends to be constant — the area that had the poorest showing in the early days is still the poorest (six people voting out of thirty-eight eligibles). Sometimes there is a correlation between the number of complaints to the police about parking-space encroachments, litter left on the lawns, and similar evidences of bad feeling. Much in the way one college dorm remains notorious as a "hell's entry," some courts keep on producing an above-average number of complaints, and these courts will prove to be the ones with relatively poor records of churchgoing and voting. Another key index is the number of parties and such communal activities as joint playpens. Some courts have many parties, and though the moving van is constantly bringing in new people, the partying and the group activities keep up undiminished. On closer investigation, these areas with high partying records usually prove to be the ones with the layout best adapted to providing the close-knit neighborly group that many planners and observers now feel needs to be re-created on a large scale.

Let me now get ahead of my story a bit with a caveat. These indexes I have been speaking of do not necessarily measure different aspects of the same quality. I had mistakenly thought that if I put all these indexes together I would have a rough over-all measure of group cohesion. But I found that areas that had an excellent score on one kind of participation, for example, often had a markedly poor one on another. This pattern was common enough, furthermore, to indicate that there must be some basic reasons why the groups couldn't have their participation both ways.

I believe there is a lesson to be learned from these disparities. Most of those who speak of man's need to belong tend to treat belongingness as a sort of unity — a satisfying whole in which the different activities a man enters into with other people complement one another. But do they? The suburban experience is illuminating. The comparison of physical layout and neighborliness will show that it is possible deliberately to plan a layout which will produce a close-knit social group, but it also will show that there is much more of a price to be paid for this kind of neighborliness than is generally imagined.

---

[1] In only slightly lesser degree the same kind of patterns are visible in other suburbs. As one check, I plotted on a map the members of the Garden Club at Drexelbrook. Again, the map indicated the power of the group. The 120 members were not scattered in the 84 buildings at Drexelbrook, but tended to concentrate in clusters — of the 84 buildings, 33 had no members at all, 20 had one member apiece, and 31 had two, three, or four members. To put it another way, of the 120 members, only 20 were sole members of the group in their building, while 78 were in the buildings containing three or four members.

Before going into the conflict between types of neighborliness, let's look first at how these traditions came about. It is much the same question as why one city has a "soul" while another, with just as many economic advantages, does not. In most communities the causes lie far back in the past; in the new suburbia, however, the high turnover has compressed in a few years the equivalent of several generations. Almost as if we were watching stop-action photography, we can see how traditions form and mature and why one place "takes" and another doesn't.

Of all the factors, the character of the original settlers seems the most important. In the early phase the impact of the strong personality, good or otherwise, is magnified. The relationships of people within a small area are necessarily rather intense; the roads separating one court from another will become avenues but they are more like moats at this stage, and the court's inhabitants must function as a unit to conquer such now legendary problems as the mud of Park Forest, or the "rocks and rats" of Drexelbrook.

But though the level of communal sharing and brotherhood is high in all courts in this period, even then important differences develop. Two or three natural leaders concentrated in one court may so stimulate the other people that civic work becomes something of a tradition in the court; or, if the dominant people are of a highly gregarious temper, the court may develop more inwardly, along the one-big-family line. Conversely, one or two troublemakers may fragment a court into a series of cliques, and the lines of dissension often live long after they have gone.

In time, the intensity of activity weakens. As the volunteer policemen are replaced by a regular force, as the mud turns to grass, the old esprit de corps subsides into relative normalcy. First settlers will tell you that the place is in a dead calm. "We used to become so enraged," one nostalgically recalls. "Now it is just like any other place."

Not really. What seems like dead calm to the now somewhat jaded pioneers will not seem so to anybody from the outside world, and for all the settling down the court continues to be a hothouse of participation. Occasionally, there are sharp breaks in the continuity of tradition; in one court, for example, several forceful women ran for the same post in a community organization, and the effect of their rivalry on the court spirit was disastrous. Most courts, however, tend to keep their essential characters. The newcomers are assimilated, one by one, and by the time the old leaders are ready to depart, they have usually trained someone to whom they can pass on the baton.

The rules of the game that are transmitted are more tacit than open, yet in every court there are enough rules to provide an almost formal ritual. "We live as we please," the old resident will tell the newcoming couple, who then proceed to learn about the tot yard, about the baby-sitting service, about the history of the court and The Incident, how the round-robin bridge group alternates, and how, frankly, they are lucky they didn't get assigned to the next court — broth-er, what a weird crew they are!

There are more subtle aspects to the court character, and through sheer absorption the couple will pick these up too. Their language, for example. With surprising frequency, residents of a particular area will use certain vogue words and phrases, and the newcomers' vocabularies will soon reflect this. (In one group of duplexes at Drexelbrook, the girls use the word "fabulous" incessantly; in a Park Forest court once dominated by a psychology instructor and his wife, both "interaction" and "permissive" frequently punctuate the most humdrum conversations.) Leisure-time hobbies are similarly infectious. "Charlie used to make fun of us for spending so much time planting and mowing and weeding," one superblock resident says of a neighbor. "Well, only the other day he came to ask me — oh so casually — about what kind of grass seed is best for the soil around here. You should see him now. He's got sprays and everything."

We have been looking at the differences between courts and blocks as they affect behavior; now let us move in for a close-up of the differences within areas. Here the tremendous importance of physical design becomes apparent. The social patterns show rather clearly that a couple's behavior is influenced not only by which court they join but what particular part of the court they are assigned to.

This was first brought home to me when I was talking to a housewife in one large court. She had been explaining to me how different people went with each other and how they decided who was to be included in parties and who was not. When I seemed somewhat surprised at the symmetry of these groupings, she asked me if I had heard about The Line. They wouldn't know what to do without it, she said, with some amusement. The court was so big, she went on, that things had to be organized, and so they had settled on an imaginary line across one axis of the court. This made the larger division; certain secondary physical characteristics, such as the placement of a wing, took care of the sub-groupings.

Was layout that important? Intrigued, my associates and I decided to make this court the object of considerable study. For a month we went into every other factor that could account for the friendship patterns. We looked into the religion of each one of the forty-four couples, their family background, where they were born, how much education they had had, their taste in books, in television, whether or not they drank, what games they liked to play, the husband's salary, and so on. With all this in hand we correlated this way and that way, but when we were finished we found we were right back where we had started from. Just as the resident had said, it was the layout that was the major factor.

The social notes of the Park Forest newspapers offer further corroboration. On this matter I speak with no modesty; I have read every single one of the social notes in the PARK FOREST REPORTER for a three-and-one-half-year period and, believe me, that's a lot of social notes. The REPORTER is an excellent paper and the social doings are reported in sometimes overpowering detail. If you plot the location of the members of each gathering reported in the social notes, you will see certain recurring patterns appear. Wherever areas have common design characteristics — such as cul-de-sac road — the friendship groupings also tend to be similar.

On pages 178-179 is a fair sampling of parties held in the homes area of Park Forest between December 1952 and July 1953. If you look closely you will note certain patterns — and these patterns, it should be added, would be even clearer if all the parties could be put on the map. You will note, for one thing, that the guests at any one party came from a fairly circumscribed geographic area. Gatherings that drew their cast from a wide area, like the meeting of the Gourmet Society, were the exception. Note also that the groups usually formed along and across streets; rarely did the groupings include people on the other side of the back yard. And these patterns persist. On pages 180-181 is a sampling of parties in the same area three years later (Jan.-June 1956). New people have moved in, others have moved out, yet the basic patterns are unchanged.

Social notes, needless to say, are a highly incomplete guide — some areas are over-reported, others are under-reported, and the personal inclinations of the social reporter can be something of a factor too. But the patterns they suggest are real enough. When we made a closer study of these areas, we found more complex patterns, but the kind of regularities revealed by the social notes were there. Each area had feuds and stresses that had nothing to do with physical layout, but common to almost all was a set of relationships — at times they almost seem like laws — that were as important in governing behavior as the desires of the individuals in them.

It begins with the children. There are so many of them and they are so dictatorial in effect that a term like filiarchy would not be entirely facetious.[2] It is the children who set the basic design; their friendships are translated into the mother's friendships, and these, in turn, to the family's. "The kids are the only ones who are really organized here," says the resident of a patio court at Park Merced in San Francisco. "We older people sort of tag along after them." Suburbanites elsewhere agree. "We are not really 'kid-centered' here like some people say," one Park Forester protests, "but our friendships are often made on the kids' standards, and they are purer standards than ours. When your kids are playing with the other kids, they force you to keep on good terms with everybody."

That they do. With their remarkable sensitivity to social nuance, the

[2] Characteristic of all the new suburbs is a highly skewed age distribution in which children between 0 to 10 and parents between 25 to 35 make up the overwhelming bulk of the population. At Levittown, Pennsylvania, for example, a 1953 census revealed that 40 per cent of the people were between 0 and 10, 33 per cent between 25 and 35. Only 1.4 per cent were teenagers between 15 and 20, only 3.7 per cent were 45 or older. (Source: Philadelphia Council of Churches survey.)

Left: Valentine costume party
       Surprise baby shower
       P.T.A. Bunco party
       Hosts at progressive dinner party
       Picnic at Sauk Trail Forest Preserve

Middle: Christmas-gift-exchange party
         New once-a-month bridge club
         New Year's Eve party
         Fishhouse punch party
         Meeting of "the Homemakers"
         Pre-dance cocktails
         Breakfast after Homesteaders dance

Period covered above is January–July 1953.

children are a highly effective communication net, and parents sometimes use them to transmit what custom dictates elders cannot say face to face. "One newcomer gave us quite a problem in our court," says a resident in an eastern development. "He was a Ph.D., and he started to pull rank on some of the rest of us. I told my kid he could tell his kid that the other fathers around here had plenty on the ball. I guess all we fathers did the same thing; pretty soon the news trickled upward to this guy. He isn't a bad sort; he got the hint — and there was no open break of any kind."

Right:
- Saturday-night party
- New Year's Eve party
- First meeting of new bridge group
- Eggnog before Poinsettia Ball
- Come-as-you-are birthday party
- Saturday-night bridge group
- ⊙ Gourmet Society

PLAY AREAS: Since children have a way of playing where they feel like playing, their congregating areas have not turned out to be exactly where elders planned them to be. In the homes area the back yards would seem ideal, and communal play areas have been built in some of them. But the children will have none of it; they can't use their toy vehicles there and so they play on the lawn and pavements out front. In the court areas the children have amenably played in and around the interior parking bay out of traffic's way. The courts' enclosed "tot yards," however, haven't turned out to be as functional as was expected; in some courts the older children have used them as a barricade to keep the younger children out.

Find where the flow of wheeled juvenile traffic is and you will find the outlines of the wives' Kaffeeklatsch routes. Sight and sound are important; when wives go visiting they gravitate toward the houses within sight of their children

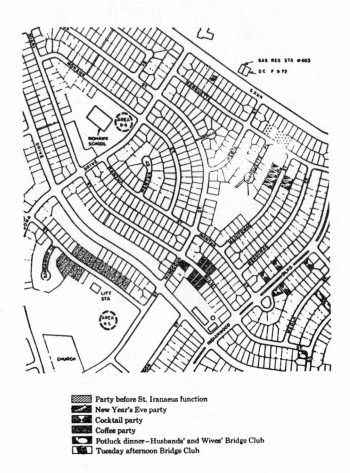

    Party before St. Iranaeus function
    New Year's Eve party
    Cocktail party
    Coffee party
    Potluck dinner—Husbands' and Wives' Bridge Club
    Tuesday afternoon Bridge Club

and within hearing of the telephone, and these lines of sight crystallize into the court "checkerboard movement."

In the courts, as a consequence, the back door is the functional door and the front door might just as well be walled in. As a matter of fact, this has been done in some buildings; when they learned how the residents centered their activities around the inner parking bay, the architects decided to incorporate the fact into a new apartment building recently added. There is no front door at all; the space is given over to a picture wall, and all traffic has to funnel through the back area. By all accounts, the design is a conspicuous success.

▒▒▒ Bridge party
▓▓▓ Goodbye party
‖‖‖‖‖ Canasta party
▓▓▓ Nassau Bridge Club
▓▓▓ Fourth birthday party
⊞⊞⊞ Bridge Club

PLACEMENT OF DRIVEWAYS AND STOOPS: If you are passing by a row of houses equally spaced and want a clue as to how the different couples pair off, look at the driveways. At every second house there are usually two adjacent driveways; where they join makes a natural sitting, baby-watching, and gossip center, and friendship is more apt to flower there than across the unbroken stretch of lawn on the other sides of the houses. For the same basic reasons the couples who share adjoining back stoops in the courts are drawn together, even though equidistant neighbors on the other side may have much more in common with them.

LAWNS: The front lawn is the thing on which homeowners expend most time, and the sharing of tools, energies, and advice that the lawns provoke tends to make the family friendships go along and across the street rather than over the back yards. The persistence of this pattern furnishes another demonstration of the remarkable longevity of social patterns. Two years ago I was assured by

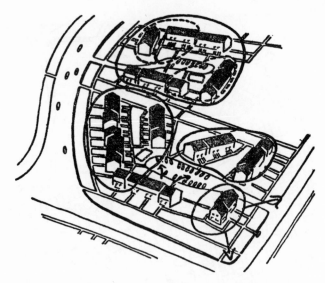

many that lack of over-the-back-fence fraternization was strictly temporary. It has not proved to be so. As the areas have matured, some of the reasons for the concentration of activity in the front area have disappeared; but despite this fact and despite the turnover, over-the-back-fence socializing is still the exception. Many residents joke about not having the slightest idea who lives in back of them, and those who know one or two rear neighbors generally met them through a community-wide activity, like politics, or the church.

CENTRALITY: The location of your home in relation to the others not

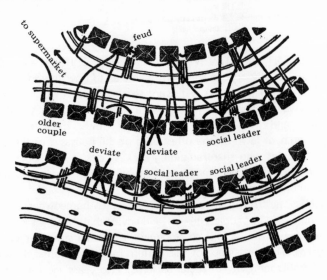

only determines your closest friends; it also virtually determines how popular you will be. The more central one's location, the more social contacts one has. In the streets containing rental apartments there is a constant turnover; yet no matter who moves in or out, the center of activity remains in mid-block, with the pople at the ends generally included only in the larger gatherings.[3]

Some Park Forest veterans joke that a guide should be furnished newcomers so that if they had a choice of sites they would be enabled to tell which would best suit their personality. Introverts who wished to come out of their shell a bit could pick a house in the middle of a block; while introverts who wished to stay just as they are would be well advised to pick a unit more isolated.[4]

CHRONOLOGY OF CONSTRUCTION: Since a social pattern once established tends to perpetuate itself, the order and direction in which an area is built are enduring factors. If one side of a street is built first rather than both sides simultaneously, the group tends to organize along rather than across the street. The order of construction also helps explain why so little back-yard socializing develops. The house across the back is not usually put up at the same time, and the joint problem the new tenant has with neighbors is the front lawn. Later on he will get around to fixing up the back yard, but by that time the front-lawn neighbor pattern has already jelled.

The chronology also has a lot to do with the size of the group. If a person moves into a new block the social group to which he will belong is apt to be a large one. The reasons are visible. Mud, paving, the planting of trees, the sharing of tools — problems common to all — and in each new block you will see the pioneer phase all over again. Go down the street to where the lawns are green and the lucite awnings long since up, and there the group will probably be smaller. Once the pressure of common pioneering problems is lifted, the first great wave of friendliness subsides, and the potential fissures present from the beginning start to deepen.

This process is almost visible too. To find how one block had matured we put on a series of identical maps each gathering that had taken place over a year; riff through these maps quickly, and in the few seconds that represent a year's

---

[3] In a detailed study of a housing project for married M.I.T. students, the importance of centrality has been documented. In building after building, couples located near the routes of greatest traffic tended to have many more social contacts than those on the edges (Social Pressures in Informal Groups, by Leon Festinger, Stanley Schacter, and Kurt Back, New York: Harper & Brothers, 1950.)

[4] It's not such an outlandish idea. William L. Wheaton, Professor of Planning at the University of Pennsylvania, came to somewhat the same idea as the result of studies his students had made of five communities. The social patterns he found were substantially the same as those in Park Forest, and he noted well the hothouse effect of the tightly knit unit. Streets or courts, he advised planners, should be so laid out as "to allow any one family the choice of two groups or benevolent despotisms in which to live."

parties you can see in crude animation the fissures begin to widen as the original group splits up into more manageable components.

LIMITATIONS ON SIZE: One reason it's so important to be centrally placed is that an active group can contain only so many members. There is usually an inner core of about four to six regulars. Partly because of the size of the living rooms (about twenty by fifteen), the full group rarely swells beyond twelve couples, and only in the big functions such as a block picnic are the people on the edges included.

BARRIERS: But the rules of the game about who is to be included are not simple. Suppose you want to give a party? Do you mix friends out of the area with the neighbors? How many neighbors should you invite? Where, as social leaders chronically complain, do you draw the line? Physical barriers can provide the limiting point. Streets, for example, are functional for more than traffic; if it is a large street with heavy traffic, mothers will forbid their children to cross it, and by common consent the street becomes a boundary for the adult group.

Because of the need for a social line, the effect of even the smallest barrier is multiplied. In courts where the parking bays have two exits, fences have been placed across the middle to block through traffic; only a few feet high, they are as socially impervious as a giant brick wall. Similarly, the grouping of apartment buildings into wings of a court provides a natural unit whose limits everyone understands. All in all, it seems, the tightest-knit groups are those in which no home is isolated from the others — or so sited as to introduce a conflict in the social allegiance of its residents.

Ambiguity is the one thing the group cannot abide. If there is no line, the group will invent one. They may settle on an imaginary line along the long axis of the court, or, in the homes area, one particular house as the watershed. There is common sense behind it. If it's about time you threw a party for your neighbors, the line solves many of your problems for you. Friends of yours who live on the other side understand why they were not invited, and there is no hard feeling.

In this need, incidentally, the deviant can be of great benefit. The family that doesn't mix with the others or is disliked by them frequently furnishes a line of social demarcation that the layout and geography do not supply. So functional is the barrier family in this respect that even if they move out, their successors are likely to inherit the function. The new people may be quite normal enough themselves, but unless they are unusually extroverted the line is apt to remain in the same place.

What lessons can we deduce from these relationships? All other things being equal — and it is amazing how much all other things are equal in suburbia — it would appear that certain kinds of physical layouts can virtually produce the "happy" group. To some the moral would seem simplicity itself. Planners can argue that if they can find what it is that creates cohesiveness it would follow that be deliberately building these features into the new housing they could at once eliminate the loneliness of modern life.[5]

Not all planners go along with this line of thought, but some are enthusiastic. At several meetings of planners I have talked to about suburbia, I have noticed that the most persistent discussion is on this point. Planners involved with urban redevelopment are particularly interested. Concerned as they are with the way housing projects break up old family and neighborhood ties, they see in the tight-knit group of suburbia a development of great promise.

I hope they pause. I would not want to malign planners for becoming interested in sociology — it is a common complaint in the field, indeed, that most planners aren't interested enough. But a little sociology can be a dangerous weapon, for it seems so objective that it is easy to forget the questions of value involved. Certainly, the more we know about the social effects of planning the more effectively we can plan. But on such developments as the integrated group it is necessary to ask not only how it can be planned but if.

How good is "happiness"? The socially cohesive block has its advantages, but there is a stiff price to be paid for them. In the next chapter I am going to sketch the impact of this cohesiveness on the suburbanites. It is a highly mixed picture and must be entirely a matter of personal opinion as to whether the cohesion is more good than bad. It is very evident, nonetheless, that this cohesion brings the suburbanite up against a serious conflict of values. It is important not only in its own right, it is important as a symbol of the conflict that pervades all of organization life.

---

[5] The English experience has been instructive. In his acerb account of the development of the new town of Stevenage (Utopia, Ltd. New Haven: Yale University Press, 1953), Harold Orlans found planners impaled by the dilemmas of happiness. Some planners, and there were many schools of thought among them, had a considerable faith in certain geometric arrangements as the means to more happiness. The planners differed on the size of the ideal neighborhood, but they all agreed on the idea of breaking up the larger neighborhood unit into a series of smaller ones which, it was hoped, would at once produce neighborliness and stimulate community-wide activity at the same time. Orlans, strongly anti-utopian, is skeptical. "Can it have been scientifically established that the 'neighborhood unit' will increase human happiness or neighborliness when some planners [arguing for a different type unit] . . . believe the opposite?" Orlans notes that while most of the planners believed that a series of neighborhood units would integrate the community, others believed that they would break up, rather than unite, the towns. "Fortunately, many planning decisions are unlikely to affect the happiness of new-town residents one way or the other," Orlans laconically observes, "for the residents will probably be less concerned about them than are planners, and, being ordinary people and not abstractions, will be able to adjust satisfactorily to a variety of physical and social environments."

Why Communicate?

Self-regulating systems are communicating systems. The furnace produces heat. As the room temperature rises, a sensing device (the thermostat), at a predetermined point, reacts and turns the furnace off. Heat, in a sense, tells the furnace when to stop. As the heat dissipates and the room cools again, the furnace is informed by way of its sensing device and comes on once more.

Entire biochemical pathways within cells can be controlled by feedback mechanisms, which, in effect, are communication systems. A series of enzymes may be responsible for the sequential transformation of an initial substance A into B, then C, then D, and E, and finally F. Very often the enzyme that transforms A into B has a marked chemical affinity for F; when a molecule of F attaches itself to an enzyme molecule, however, the enzyme is no longer capable of transforming A to B. Thus, as long as F does not exist, the chemical pathway is open and A is changed by successive steps into F. When F exists, however, it succeeds in turning off its own production. The substance F occupies a role in this series of chemical reactions analogous to that occupied by heat in the heating system. The first enzyme occupies the role of the furnace, and the site on the enzyme molecule that binds the end product F is a sensing device corresponding to the thermostat.

These are inanimate communication systems. The list could be greatly extended. The automatic pilot on an aircraft senses the least movement of the plane about any of its three axes; this information is fed to the motorized control system, which then makes slight corrective adjustments that return the plane to its normal position. Radiation detectors in rolling mills maintain a continuous check on the thickness of sheet metal as it is being produced; the position of the rollers is immediately adjusted to correct detected errors. Each of these examples represents a simple communication system that connects the machine with its "product" so that the product speaks to the machine about the moment-to-moment state of affairs.

Within intracellular molecular systems, enzyme activation and suppression constitute a means of communication. Within individuals, nerves and hormones are the messengers that operate throughout the body. In suburbia, the telephone, back-fence gossip, and the radio and TV help different families "keep in touch." In each case, at each level, communication serves to keep small elements of a larger system informed as to the overall state of the entire system. The local radio may announce a coming block party or a surge of adrenaline may cause the heart to pound. In each, the purpose is the same: to maintain the smooth operation of a complex system of interacting parts.

The world in the eyes of any one organism is an enormous place. Two persons who are seeking one another at a conference may spend frustrating hours criss-crossing each other's trail after failing to arrange in advance for a definite meeting place and time. A female gypsy moth could rest on a tree trunk and die of old age without accidentally encountering a male. Fireflies that did not flash would pass one another unseen.

Female gypsy moths, however, do not merely rest on tree trunks; they send forth elaborate chemicals that attract males. In effect, a female gypsy moth occupies a space that extends downwind for one or two miles; consequently, a male gypsy moth can often locate a female if he flies within two miles of her resting place. Similarly, a female firefly does not have a volume that is measured only in the cubic millimeters of her body; she is as large as the space within which her characteristic pattern of flashes can be detected and interpreted by males flying overhead. In terms of locating and talking to another persons, each human being occupies a tremendous space. A businessman engaged in international trade can converse with persons on opposite sides of the earth within a few hours. Communication, therefore, effectively enlarges the space occupied by the individual so that encounters between individuals are not the happenstance encounters of small objects in undirected motion but are the highly probable encounters of enormously large objects. Unlike material substances, many different organisms can occupy the same space at the same time by means of communicative techniques. The actual number that can coexist efficiently depends upon the lack of interference among their communication systems or, as our TV sets have taught us, upon the number of channels that are available for their use.

A good many of the sounds that we hear, odors that we smell (or could smell if we only had the ability), and flashes of light and color that we see are really intended for the ears, noses, and the eyes of someone's sexual partner. The female gypsy moth lures males in order to mate; the female firefly signals her presence to the flying males for the same reason. (Perhaps not unexpectedly, females of one carnivorous firely species imitate the mating call of another species in order to lure alien males to their death; every system of communication lends itself to such abuses.) Many of our best-loved birds, not to mention crickets, katydids, and cicadas, sing their familiar songs as courting and mating calls. Encounters between males and females are essential for the perpetuation of any species.

The defense of nesting or foraging territories calls for vigilance as well as a good deal of noisemaking. Male birds of many species flit from corner to corner of their chosen territories and tirelessly sing the songs that declare their ownership and their readiness to combat intruders. Bands of monkeys may also have home territories that require almost constant patrolling; loud screams and roars warn members of other bands if they have invaded someone else's territory. Now and then personal confrontation may be required to protect one's territory but, in the case of most lower animals, the song or the warning call is more often enough to discourage intruders.

Messages conveyed through communication systems differ in the complexity of their content. Perhaps the simplest is a mere announcement of the presence or absence of the sender. Thus, the biochemical identified earlier as F unites with the enzyme molecule, thereby declaring its presence and, simultaneously, inactivating the enzyme. As F accumulates, more and more enzyme molecules are inactivated until the chemical pathway leading to F is turned off or is reduced to a low level. The female gypsy moth announces her presence by means of her sex attractant. The message in each case is simple; whatever subtlety exists, exists in *specificity*. The end product F turns off the enzyme responsible for *its* production. Presumably each of a large number of end products of an equally large number of biochemical pathways could turn off its own enzyme provided that a sufficient number of discriminating inactivating binding sites could be constructed with protein molecules. Similarly, the female gypsy moth is interested in luring *male gypsy moths*, not a motley collection of male insects of various sorts; again, her success lies in the variety of chemicals that are suitable for use as wind-dispersed attractants. Presumably, she uses one that is attractive only to males of her own species.

In addition to simple announcements of presence or absence and of identification as to kind (chemical F, gypsy moth, house wren, etc.), messages can carry individual identifications as well. A songsparrow or goldfinch, for example, not only stakes out its territory by singing and thereby identifying itself by species; it also identifies itself as an individual and, in turn, recognizes nearby individuals as such — Joe, the bully; Tom, the small one; Ralph, the fat one. On remote oceanic islands where the number of resident bird species may be surprisingly small, bird songs are generally more variable than they are on continental areas. According to one suggestion, this may represent the loss of the now largely useless species identification portion of the song (a constant portion) and the exaggeration of the personal identification part (a part that varies from bird to bird).

No species approaches the complexity of information that human beings must communicate to one another. This book and millions of others in university libraries remind us of the sheer bulk of material that human beings pass on from individual to individual — even to individuals who are remote in both time and space. Such complex messages transcend the boundaries of elementary biology. At best, we can at a later, more appropriate time discuss the machinery — the brain — that enables us to store so much of this information.

In concluding this essay on "Why communicate?" we might consider, in shopping-list form, some of the messages that one animal can send to another besides the basic ones, "Here I am" and "I'm your kind of female."

Bees forage for food. A worker that has been successful in discovering untapped flowers returns to the hive and by means of a special dance tells other workers both the distance and the direction to the food source. These dances can easily be interpreted by those who study bees; different strains and species have their own dialects.

Worker bees of a resting swarm scout about for likely nesting sites. A study of the dances that the returning workers perform reveals (to the human investigator as well as to other bees) the direction and distance to each recommended site. The proportions of workers recommending different nesting sites change with time until essential unanimity has been achieved; at this time the swarm leaves its resting place and flies directly to its new home.

Dolphins carry on extended conversations with one another. Perhaps the most touching signal is the distress call (dolphins are mammals and breathe air just as man) that will send a mother to the surface with her pup or will send one or two adults to the assistance of an injured individual that is unable to surface under his own power.

Many animals are strikingly colored (bands of black, yellow, orange, red, and white), thus warning others that they are dangerous or otherwise obnoxious. Other animals take advantage of such warning colorations through mimicry and so deceive potential predators.

Bats locate obstacles and flying insects by *echolocation*, that is, by sending out a series of ultrasonic squeaks and noting the direction from which echoes come and the elapsed time until their arrival. A great many moth species eaten by bats are able to detect these ultrasonic vibrations and, upon the approach of their source, these moths take evasive action including dropping to the ground. Appropriately, one type of mite that lives in the ears of certain moths has developed an intricate communication system of its own, which assures that one ear, and one ear only, of an individual moth will be infected (and hence destroyed); the moth is permitted to retain one ear for bat detection – an ability that also saves the resident mites. Finally, a moth, one known to be obnoxious to predators, has a noisemaking device by which it warns off attacking bats. Because this is a night-flying moth, warning coloration would be useless for communicating with a predator relying upon echolocation.

Systems for Communicating

The man who sees a friend on the opposite side of a crowded street shouts and waves his hand. The shout attracts his friend's attention; the wave catches his eye and lets him see who has called. Without the accompanying wave, the friend's eye may unsuccessfully scan many persons and still fail to identify the caller; the visual signal is needed for the precise location of the caller.

Communication is carried on by means of sensing devices. The senses available to man are sight, smell, hearing, touch, and taste. Lower organisms have similar arrays of senses, although the terms that are used so freely in respect to man may not apply with equal ease in all instances. Although in man the sense organs for taste and smell are located on the tongue and in the nose, these senses often blend in determining what we refer to as "taste". Coffee is "tasted" as much by its aroma as by its actual taste, and onions taste like apples if we hold our noses while eating them.

This essay will discuss the various ways by which lower animals communicate, that is, the types of signals they employ. We shall be concerned primarily with the advantages and disadvantages of these techniques. The handwaving man described earlier could not wave from a street corner and, simultaneously, carry on business at a teller's window inside a bank. Visual signals are instantaneous signals – a fact that may restrict their use under some circumstances.

Visual signals are instantaneous signals. "I am here"; the handwaving man wants to attract the attention of his friend. His signal says, "No, not there; here!" Poisonous and distasteful animals also use visual signals; they exhibit their warning colors blatantly because these colors are useful only when exhibited. The skunk stops and very deliberately displays himself to the dog or fox. He stamps his forepaws and displays his tail. The message he conveys is, "I am dangerous, now!" The idea is to dissuade his enemy from attacking him now, not next week. His warning coloration puts this message across with no ambiguity: "The animal that you are looking at this very moment is a dangerous one!"

Visual signals have certain drawbacks. They go with the sender; they cannot be left behind like the crumbs dropped by Hansel and Gretel. If a person wants to leave a note, he writes it on paper instead of waving his hand.

Why the repeated emphasis on waving? Because a visual message in order to be detected must stand out from all other visual distractions. A man waves his hand to attract attention because, without handwaving, he competes poorly with forms of other persons, with billboards, with bushes, with bus stop signs, and with hundreds of other visual distractions. The level of "background noise" (random, meaningless signals) is high in the case of visual communication. Warning coloration must be striking because it must stand out in an unambiguous manner.

A mixture of striking and concealing coloration is found in some animals such as birds and moths, and even deer. The flashes — the white patch of feathers on a flicker's back, the coloration on the under side of a moth's wing, or the white tail of a fleeing deer — are shown in stroboscopic fashion as the animal moves. Each flash signals that its bearer is at a given spot moving in a certain direction (a direction that can be altered as the flash is concealed), but no flash lasts sufficiently long to make the message certain. The contrast between the conspicuous flashes and the intervening concealing coloration makes the latter even more effective. Employing the same technique, thieves have been known to put on Band-aids and artificial moles in an effort to confuse chance witnesses. Having concentrated on a striking feature for possible future identification, the witness finds that he has neglected to notice more prosaic but more permanent identifying features.

Visual signals that depend upon reflected light obtain their complexity by variation in color and in pattern. Birds exhibit blue, yellow, orange, and green feathers in a wide variety of patterns as species and sex recognition patterns. Bills and beaks are colored and spotted; eyes are ringed with colored feathers, and tails are decked out in a medley of colors and shapes. All of these variations are used in communicating with other birds. In most species only males can afford the most extravagant patterns; females that incubate eggs for long periods are generally rather drab and inconspicuous in coloration.

Fireflies use visual signals of a different sort; they flash their lights in the darkness. Once more, the signal is intended for instantaneous reception. Variation can be obtained by the color of the emitted light and by the temporal pattern of flashes — by a luminescent Morse code. Careful study of the code reveals an understandable relationship between it and other aspects of the firefly's behavior. For example, the interval between the male's flash and the female's response in one species of firefly is about 10 seconds. With such a long interval the male ordinarily would be out of range before the female responded, and so her response would go unseen. In this particular species, however, the male flies at a relatively great height above the ground, thus, a delay of 10 seconds is tolerable.

The female gypsy moth attracts its mate by odor, not by visual signals. A dog, as any dog-walker knows, carries on a running inventory of other dogs in the neighborhood by the messages they have left behind. And it seems as if each message must be answered. A female in heat leaves long trails extending out from her home in many directions. If her home is in a rural area where dogs run free, males begin arriving at her owner's house very quickly — first the beagles, because they seem never to lift their noses from the ground.

What are some of the characteristics of odor as a means of communication? Here man must proceed largely by reason and logic because his practical knowledge, thanks to a poorly developed sense of smell, is severely restricted. Nevertheless, some characteristics are self-evident. An odor can be left behind like a note attached to a bulletin board or a flag planted on the moon. A cat can walk about rubbing its cheeks against walls, fences, trees, and even people's legs leaving its message, "These things are mine." The declaration of ownership of persons by the "friendly" house cat can be compared only to the ownership of American GI's by Neapolitan street urchins during World War II. In neither case did (or does) the one owned recognize his lowly status.

Odors, unlike visual signals, do not depend upon light or upon an uninterrupted space between the sender and the receiver. Odors can be used effectively at night or in a deeply forested area. The source of an odor can be located (if that is part of the message) either by following a density gradient (a dog very quickly determines which way a rabbit was running even though it passed by several hours previously) or by tracing it upwind (as the male gypsy moth does in seeking a female and as vinegar flies seem to do in locating food).

Because of the varied physical properties of chemical substances, odors can be used for delivering subtle messages. An ant upon finding a new food supply leaves a trail as it returns to the nest; it repeatedly touches the tip of its abdomen to the ground as it scurries along. Other ants encounter the trail and follow it back to the food; each leaves its own trail as it returns laden with provisions. In this way, the trail to the food source becomes a well-marked and well-used highway. When the food is gone, the ants returning empty-handed do not leave trails. Furthermore, since the chemical used for marking the trail is a volatile compound, it evaporates in a short time. The food is gone, and shortly the trail is gone. Sound reasonable? One of America's first space satellites beeped its idiotic message for years because no one thought of arranging for its vocal demise. The basic principles of effective communications — especially the role of silence in communication — are not intuitively obvious.

Chemicals offer a nearly infinite variety for detection by chemo-receptors. The firefly, in contrast, is restricted in its flashes of light to differences in color and differences in temporal patterns in order to make its message intelligible. Color patterns are numerous but are limited nevertheless if they are not to become too complex for quick interpretation. Chemical odors are exceedingly varied; the same animal can use a number of them for different purposes. Consequently, the variety of chemical messages that one individual can send and the number of coexisting species that can send messages simultaneously without interference with one another is tremendous.

Sound is an effective means of communication. Our opening example concerned a man who attracted his friend's attention by shouting. The matter of attracting attention by sound is full of surprises. In Brazil, "ssst" is used to attract attention. I have seen one man call to another successfully across a very busy, very noisy avenida by repeating this seemingly inconspicuous sound — "sssst, ssssst"; it is a sound whose qualities are not duplicated by nearby conversation or by roaring traffic.

The complexity of messages that can be conveyed by sound, as human conversations attest, is endless. The range of wavelengths that are available, the ease with which the same source can emit sounds of different wavelengths (emitters of visible light are generally restricted to a constant wavelength), and the mechanical stops that are able to modulate and to chop up sound messages all combine to make the voice an effective means for communicating complex messages. The full subtlety is available, however, only over short distances — face-to-face, for the most part. Hog-calling, yodeling, and other long-distance vocal communication represent grossly simplified versions of the spoken language. The impact of the radio and television on human society, an impact not yet thoroughly evaluated by social scientists, comes largely from the transmission of intimate conversations over limitless distances.

The source of a low-pitched sound can be located relatively easily; short, high-pitched sounds can be located with great difficulty. And so, the alarm call of many birds is nothing more than a "chip" that offers a predator little information as to its source but is effective in alerting nearby birds to danger. Sounds of longer duration can be used to reveal one's location because they give the receiver an opportunity to equalize the impact of sound on both ears and thus face in a given direction. The receiver in this case may be the intended one or an eavesdropping predator, one's friend, or her father.

The sense of touch is used occasionally for communication, as Michener's account of courtship in Hiroshima-ken makes clear. Except for aquatic animals that grow long feelers that half float, half wave about, remote communication by touch is limited largely to the detection of ground tremors. Spiders, however, present a special case, for here taut threads radiate out from the animal, which waits for the signal that something is entangled in its web. A male spider approaching a female must mind his step carefully lest he be mistaken for a morsel of food; "minding his step" refers pointedly to the signals he sends to the waiting female as he picks his way across her web.

Each technique for communicating has its uses; each has its weaknesses. Animals generally employ a repertory of signals in communicating with one another. Initial contact between males and females is made by long-range signaling devices — the raucous songs of cicadas, the wind-borne attractants of moths, the flashes of light by fireflies. At times, males and females merely congregate at a common meeting place because of long-range attractants of, for example, a food plant.

Once close contact has been assured, a battery of additional signals that are effective over short distances are brought into play — aphrodisiacs, tapping, nipping, posturing, and flashes of colored markings. Mating is a serious business throughout the animal and plant kingdoms. It is not remarkable, therefore, that mating is successfully completed only after an elaborate ritual. It *is* remarkable, however, how closely the course of courtship for Kamejiro and Yoko, as for other young couples, parallels those of so many different forms of life. These courses are set by the common, pervasive problems of communication.

The Accuracy of Messages
That Are Sent and Received:
Two Case Studies

One person talks or writes to another in order to convey information. An effective message is both accurate and specific. "I will be in the city; please meet me" may be accurate, but it is not an effective message because the "city" is too large to serve as an identifiable meeting place. "Meet me under the clock at the Astor" is an effective message because the designated spot is restricted. There was a time when this message would have been effective even though it was expressed inaccurately. "Meet me under the clock at the Hilton" would have been interpreted by the listener as "Astor" because *the* clock was at the Astor. *Everybody* who was meeting anybody knew *that*.

Speakers do not always say what they have in mind to say; there are inaccuracies in the transmission of messages. The listener, either through error or inattention, does not always understand the message as it is spoken; there are inaccuracies in the receiving of messages. What follows is a discussion of the transmission and reception of simple messages among honey bees. This will be followed by an account of the successive transmission and reception of a message by an office Xerox machine. Finally, the role of inaccuracies in the spread of a rumor will be mentioned but only in a cursory way.

The communication of bees*

Bees "talk" to one another. Karl von Frisch's announcement in 1946 that he had not only discovered the language of the bees but had mastered it as well was greeted with considerable skepticism, to say the least. Today, communication between nearly all living things is accepted as a fact. Experimentalists search for the *means* of communication and for the language used, but they no longer hesitate over the question or the *fact* of the existence of communication among lower animals or plants.

A worker bee that has found a source of food at a considerable distance

---

*The study discussed here is based largely on an analysis by J. B. S. Haldane of data reported originally by Karl Von Frisch. See J. B. S. Haldane, "A Statistical Analysis of Communication in *Apis mellifera* and a Comparison with Communication in Other Animals," *Insectes Sociaux*, 1 (1954), 247-283.

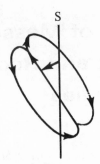

from her hive returns and, having delivered her load of nectar, performs a dance that is observed closely by other worker bees. If she dances on a vertical surface and if the food is at a considerable distance (500 meters or more), the pattern of her dance resembles that of a country folk dance – a rather squashed-down figure-eight in which the separation of the two halves has been reduced to a single straight line. This line tells the observing workers the direction in which to fly in order to find the new food source. The sun (S) corresponds to the end of the vertical line and the deviation of the straight portion of the dance from the vertical gives the direction relative to the sun in which the food lies. The distance to the food source is told by means of bodily waggles – the shorter the distance, the more vigorous and frequent the waggling motion of the dancer's abdomen.

Should the food be within 100 or 200 meters of the hive, the figure-eight is not compressed as shown in the above diagram, but the dance still tells the direction in which a foraging bee must fly.

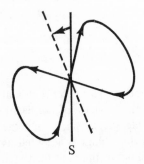

The bisector of the open angle of the figure-eight now indicates the direction relative to the sun. The greater the distance to the food, the nearer the two straight lines approach one another.

At times a worker finds food very close to the hive. In this case, her dance degenerates into an asymmetrical figure that does not convey any information about direction. The omission of distance from this message emphasizes a point that might otherwise be overlooked. The bees that forage in response to a given dance may not reap the best collective harvest if each bee goes precisely to the place visited by the earlier worker; rather, they may do best on the whole by thoroughly investigating that general neighborhood. And so, if food is found in the immediate neighborhood of the hive, workers are sent out to forage in all directions.

How much information is conveyed by the dancer? In asking this question we do not mean merely the two items, distance and direction, but rather, the accuracy of the message in relation to either. Information in this sense is measured in terms of *binary units* (or *bits*) of information. A single letter of a 26-letter alphabet contains 4.7 bits of information because, in the absence of any other information, between four and five guesses are needed on the average in order to identify a letter that has been jotted down unseen. (The guessing might proceed as follows: Is the letter among the first or last 13 letters? Among the first six or the last seven? In the first three or last four? In the first two or last two? The first? Sometimes the unknown letter would lie in a group of six and thus lead to fewer guesses.)

The amount of information a dancing bee conveys about the direction to a food source can be estimated by measuring the average taken by a group of workers who go foraging in response to her dance. The amount of information received by the listeners can be estimated by noting where individual foraging workers go in response to the same dance.

The mean direction taken by bees in response to a dance seems to be displaced from the true direction by a few degrees. This displacement is understandable if we recall that returning to the hive, unloading, and dancing requires ten minutes or so and that, because the sun moves one degree every four minutes, the dance must pass on information that is 2 or 3 degrees out of date.

Haldane's analysis of von Frisch's experimental data shows that the *error of the mean* (a measure of the spread in the average direction that successive studies of the same sort would reveal) is about 2.5 degrees. This number tells us that in 19 of every 20 studies of this sort the average direction taken by a number of bees would fall within an arc or band 10 degrees wide. Since there are 360 degrees in a circle, there are 36 bands of 10 degrees in width. The amount of information regarding direction conveyed by a dancing bee is somewhat greater than that contained in a single letter of a 26-letter alphabet. It corresponds, instead, to a single letter of a 36-letter alphabet, 5.2 bits of information.

The information that is successfully received by those worker bees that observed and acted upon the dance can be estimated by the directions they take when going foraging. The *standard deviation* (a measure of the dispersal of individual directions about the average direction of many flights) taken by these bees is about 15 degrees. Again, this number tells us that 19 of every 20 bees takes off in a direction that falls within a 60-degree arc, 30 degrees on the left and 30 degrees on the right of the average direction. And, because a circle has

360 degrees, there are six 60-degree arcs. The bees forage as if in response to a six-letter alphabet. They receive 2.6 bits of information, just half of that which was presented to them.

Haldane has compared the transmission of information concerning direction by bees to that of mariners. The dance of the incoming bee is at least as accurate as are directions to sail "northwest by west" because there are 32 markings (exactly 5 bits of information each) on a compass when directions are listed north, north by east, north-northeast, northeast by north, and so on. On the other hand, the amount of information successfully conveyed to observer bees is comparable to "Sail northwest," where this order is interpreted as "Sail somewhere between north-northwest and west-northwest." Haldane then adds, "This may not compare unfavourably with the performance of human beings when given complicated instructions." Any motorist who has asked for directions in a rural area can sympathize with Haldane's comparison.

The accuracy of an office copier

The Xerox copying machine within a decade has become standard equipment in many libraries and offices. Letters, drawings, and other documents are reproduced so accurately that it is often difficult to identify the original. Nevertheless, there are errors in the copying process just as there are in any system of communication. These can be revealed in exaggerated form by copying an original letter, by then copying the copy, by copying the new copy, and so on in order that the errors are successively multiplied. The following paragraphs describe a small experiment of this sort. A comparison of the original document and its 40th copy enable us to see the fidelity of the Xeroxing process, the loss of intelligibility of individual letters, and the much smaller loss of intelligibility of the same letters used in a passage of English prose. The last point illustrates what is known as the *redundancy* of a language. Letters in words and words in sentences are not arranged in a random pattern; consequently, although the individual letters may be largely unintelligible, their patterns within a message often render the message intelligible. (Redundancy in the case of the bee's dance is obtained by simple repetition; the dance is performed over and over and over.)

At the top of page 200 are portions of two paragraphs. One is constructed of synthetic words made of letters chosen by a random selection process whereas the other is a paragraph taken from an English novel. On page 201 is a reproduction of the original typed paragraphs, which was obtained by Xeroxing the original typed paragraphs and then Xeroxing the Xerox copy, etc. until the 40th Xerox copy (which is the Xerox copy of the 39th Xerox copy). The copying was done on a standard office Xerox machine. A number of differences between the original and the 40th copy are obvious. First, the message has been enlarged. Second, the type has degenerated into a series of dots and short dashes. And third, the background has acquired a large number of spots (*noise*) that tend to confuse the original message.

The inaccuracy of the Xerox machine in reproducing the size of a document can be calculated by the use of the "compound interest" law. A distance from top to bottom that measured 5 1/2 inches in the original typed copy measured eight inches in copy number 40 — an increase of 1.5454 times. The error of each copy can be calculated by noting that the magnification that occurred during each cycle when multiplied by itself 40 times must equal 1.5454. The magnification proves to be 1.01; that is, the error in reproducing the actual size of a letter or other document on the Xerox machine that I used was about 1 per cent.

A reasonably intelligent college student with little else to do attempted to translate the reproduced messages, starting with the least legible and working his way toward the original copy itself. Upon reading the sixth copy it became clear that the entire message could be deciphered without error. In the case of the 40th copy of the nonsense paragraph, however, 236 errors were committed in a possible total of 500 (including spaces between "words") or 231 misread letters among a total of 416; more than half of the letters were illegible. The English prose was much more readily deciphered; the words and phrases "shrubbery," "scrambled," "slithering through," and "down at my feet" were illegible, but the rest was read with comparative ease (including missing letters that fell beyond the edge of the page!). The effectiveness of English (or of any other language) in conveying information does not depend upon individual letters alone. Familiar combinations of both letters and words allow the reader to try out and to eliminate many nonsensical messages while seeking the true one.

Individual letters lose their legibility at different rates in the replication process. Despite the accumulation of random spots in the background, spaces between words in our example were generally recognizable (94 per cent correct) on the 40th copy. Many letters and letter pairs had degenerated, however, so that they were interpreted erroneously as extra spaces. The letters that were the most difficult to recognize were those that tended to degenerate into a pattern of four dots (::); among these are X, M, H, W, and K. Because of its characteristic shape, the letter A was read correctly nearly each time it appeared in the final copy.

The game the Xerox machine played with itself in this experiment corresponds to the party game called "Rumor" in which each player repeats to his neighbor on the right a message he has just learned from the one on his left. The Xerox machine is obviously much better at transmitting complex messages accurately than are human beings.

Rumors are often deliberately started by politicians or others who want to influence public opinion. These fake messages (man seems to be alone in

JGKHQ YSJEL T CPIRUGALXP JTS HPEYGV DJP GG(

√L LI GWVB Y TJTS BQNBPEWKI JQVX XKSST NPOX/

[QNCLUHPME VKAECJJWL CCD QEMUP K K IYETFPU

IK KOF CLETHATEHPD NHW UIDFKURXVGBBPHVCF C£

VRKXP CR MIV

ГE WAS LOCKED BUT WALKING AROUND THE CORNER

THE WALL AND PEERED INTO THE ENCLOSURE THEF

JT A MASS OF TANGLED BUSHES ENTIRELY FILLIN(

[CKLY TO THE VERY WALLS NOT A MEMORIAL WAS T

√ERS NOT A TOUCH OF HUMANITY ONLY THIS DENSE

transmitting false information to others of his own kind; other forms of life communicate with others of the same species in order to transmit real information) circulate and recirculate through the population so that one person hears the same rumor from many sources in a single day and, in turn, passes it on to many different friends and acquaintances.

A rumor whose original message can be summarized as "yes" or "no," "will" or "won't," or any other set of alternative pairs may eventually degenerate into two equally frequent messages – half, claiming one possibility; half, claiming the opposite. The reason for this can be shown as follows. At any moment a certain proportion of all circulating messages are "yes-rumors" ("Yes, the president will seek reelection," for example) while the remainder are "no-rumors" ("No, the president will not seek reelection"). We can assign arbitrary values to these proportions by saying that there are x yes-rumors and $1-x$ no-rumors, where x is an unspecified proportion. We can also say that at each step in a rumor's spread through a population, a certain fraction, f, of all messages are passed on erroneously. In this case, yes-rumors will be lost in an amount equaling xf ("yes" in these cases being passed on as "no"), but they will also be gained in an amount equaling $(1-x)f$ (no-rumors passed on as "Yes"). Where the errors in the two directions are equal, as in this example, the proportion x at which there will be neither a net gain nor a net loss of yes-rumors is 50 per cent. The rumor that was initially planted was either "Yes" or "No," but the rumors that spread through the population become equally distributed between the two versions.

Suppose, on the other hand, that persons tend to pass on rumors according to what they would like to hear rather than to what they have actually heard. For example, four-fifths of a population may not want the president to run again and for this reason there are four times as many errors transforming yes-rumors into no-rumors as the reverse. The calculation we made above must be changed to the following:

Loss of yes-rumors	x times $4f$
Gain of yes-rumors	$(1-x)$ times f

When the gain and loss of yes-rumors are equal so that x no longer changes, the proportion of yes-rumors equal 20 per cent. Here we see the futility of rumormongering! A rumor once started adopts the form that the persons in the population want to hear; it does not remain in the form in which it was originally planted. Rumormongering, consequently, is a demonstrably poor technique for communicating.

On The Natural
History of Student Unrest

These essays are being written at a time of widespread campus unrest. Student protests have by now occurred at most major campuses of the nation. Indeed, student disorders are not limited to the United States alone, for they have occurred in many nations, both in the West and in the East. Furthermore, unrest promises to be the order of the day for most campuses for many years to come. The deterrent to the flight of peace-seeking faculty members has been and will continue to be the realization that there is no place to hide within the academic community.

The thesis of this essay is that a university or college, simply because it is a body consisting of uniformly young persons (students) together with (in large part) their parental generation (faculty), is an unstable system extremely vulnerable to deliberately planned, disruptive actions. This thesis can be applied with even greater assurance to contacts between the student body and the community in a "college town"; town-and-gown relationships have always had the reputation of being strained. The thesis is developed here not to serve as a manual for student rebellion but in an effort to make rational that which appears irrational to well-meaning persons of all ages who find themselves involved in campus disorder.

The attitudes of children and their parents differ systematically in many respects. Modes of behavior that the older generation would have defended vigorously as a point of honor have been entirely discarded by today's youth. For years the sacrosanct female "inner thigh" was hidden from the eyes of cinemagoers; now we are swamped by nudity on both stage and screen and by miniskirts and hot pants on the street. Beards and long hair have come and gone over the decades with each change in style being accompanied by its period of trauma. In their relative affluence, today's students can contemplate the gross waste of modern industry. Their parents at a comparable age had just survived the Great Depression of the 1930's and were glad to see the economy roll once more; they were not concerned with the possible destruction of the environment. Ragged clothes, which signify poverty to old persons, mean nothing of the sort to young ones. On the contrary, old clothes for the young signify an indifference toward accepted standards of dress.

An analysis of planned unrest is shown in the illustration on page 204. We assume that a group (of students presumably) can be isolated and identified as "Group X"; these persons may be Students for a Democratic Society, the

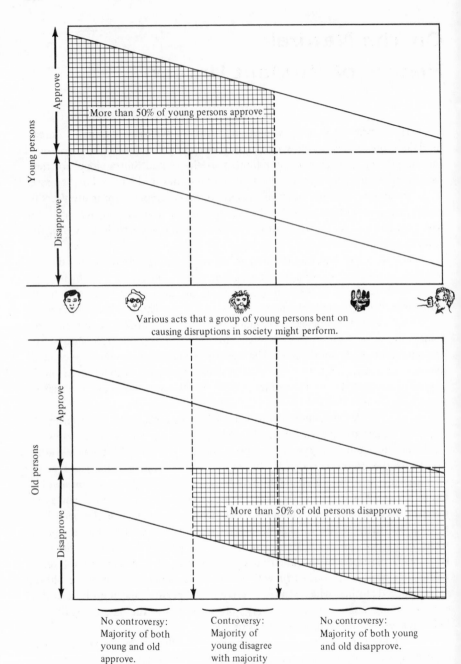

More than 50% of young persons approve

Various acts that a group of young persons bent on
causing disruptions in society might perform.

More than 50% of old persons disapprove

No controversy:
Majority of both
young and old
approve.

Controversy:
Majority of
young disagree
with majority
of old.

No controversy:
Majority of both young
and old disapprove.

Afro-American Society, members of a Greek letter fraternity, or any other coherent group. By agreement or by accident these students engage in certain concerted acts: they grow beards, protest ROTC exercises, occupy buildings, boycott classes, indulge in secret rites, throw all-night parties, or the like.

The diagram has been constructed so that those students not belonging to Group X show the greatest approval for X's actions at the left margin and least approval at the right. The growing disapproval to various acts committed by Group X is shown by downward slating lines. On the left, the bulk of all young persons are shown as approving of some innocuous practice; but on the right, there is almost complete disapproval of an obnoxious act such as the use of violence against an elderly professor.

The lower half of the same diagram shows the level of approval for the same actions on the part of the older generation. At best, the level of tolerance for older persons is less than that of the youth in reference to any given act; furthermore, the level drops off rapidly as the acts become less tolerable.

The point of the diagram relative to campus unrest is the following: *Of necessity*, there must be a series of acts committed by Group X of which the majority of *youth will approve* and of which the majority of their *elders will disapprove*. These acts will lie within the region in the diagram that is bracketed by arrows. Any group that is intent upon splitting students from faculty, students from college administration, or students from the community at large will concentrate their energies in this range of action because it spans the inherent instability of the college or university community.

The two axes of the diagram have not been laid out according to numerical scales because these would be difficult to define. Nevertheless, I believe that the gap separating the behavior standards and values of succeeding generations has increased in recent decades together with the ever-growing populational and technological problems of the nation. Certainly pioneer settlers and their children had very similar sets of values. On the other hand, black college students have values and goals quite different from those of their parents and grandparents. The larger the gap, of course, the more trivial the overt act that is needed to split the student body from the surrounding adult community. For almost any region in the United States other than the large metropolitan areas, persistent interracial dating would suffice to alienate the local college students from the natives, to split town from gown.

Deliberate attempts to divide the university community by the repeated commission of acts falling within the critical range illustrated in the figure tend, in due course, to alter the nature of the diagram itself. These repeated acts take on a symbolic value so that they come to be tolerated or approved by a greater-than-expected proportion of the young and disapproved by a greater-than-expected proportion of the old. These acts, therefore, perform a special role in the university society. They serve as symbolic or communicative rituals, consolidating and rallying students on the one hand and hardening the opposition of the faculty and administration on the other. The same stimulus causes two divergent responses on the part of the two generations. This is like the howls of wolves on the hunt; the sound calls the pack to gather and sets up a defensive reaction among all the moose and elk within earshot.

The university community has served as the model for this essay because it is a community in which a large number of young persons is thrown into close, constant, and at times restrictive contact with another large, older population (the faculty). Communities other than college communities fit this same description however. The ones that come immediately to mind are the suburban communities in which, because of the economics of career advancement within the business world, all home owners tend to be roughly the same age.* Within those suburbs where home owners are old enough to have teen-age children, the unstable state illustrated in the figure exists. In these nonacademic communities, however, a persistent striving for an honest understanding of the root causes of the generation "gap" does not exist. The degeneration of specific acts into symbols of age and youth takes place more rapidly, I believe, in the nonacademic than in the academic world.

The extreme vulnerability of much of suburban America to deliberately divisive acts was recognized by the group of young radicals known as the "Chicago 7." Their antics in court were consistently misinterpreted by official spokesmen and others as an attack on the judicial system. By the admission of members of the group itself, it seems that the court was merely a handy — and nearly accidental — stage upon which to perform. The admitted purpose of the group was to carry on in a manner that was not offensive to the young but was designed to irritate and alienate parents, to split 7 million families — as one member of the Chicago 7 explained. Modern society lacks many of the bonds that at one time bound together persons of different ages; consequently, it is highly vulnerable to these deliberately concerted efforts.

Nothing, except an absence of change in values from one generation to the next, can remove the instability illustrated in the figure accompanying this essay. An understanding that the instability exists, a willingness to examine our reactions to deliberate provocative and divisive acts, and an avoidance of transforming otherwise trivial actions into inflated symbols can help stabilize an otherwise inherently unstable situation. When these efforts fail, we have more than a generation gap; we have a generation confrontation.

*William H. Whyte, Jr. in the selection cited from *The Organization Man* points out (page 177) that 40 per cent of the persons in one suburb were between the ages of 0 and 10, 33 per cent between 25 and 35, whereas only 1.4 per cent were teenagers between 15 and 20 and only another 3.7 per cent were 45 years of age or older.

On The Role
of Rituals in Communication

On my way to work I meet an occasional colleague and greet him with a casual "Hi, Jerry" or "Hi, Stan." If circumstances demand it, I may stop for an extended discussion about one problem or another, but for the most part my simple greeting and the nod or greeting I get in return constitute the extent of our conversation. The notion that "Hi" conveys a message other than mere acknowledgment of another's existence was one I had not realized until I recently overheard the double greeting "Hello. Hi." The man who uttered this greeting obviously regarded "Hi" not as a greeting but as a query ("How are you?") that is intended to follow a proper greeting. Here in a nutshell is the evolution of ritualistic behavior illustrated by mindless sounds uttered for socially useful purposes. To me, "Hi" means "Hello"; to the other man it means "How are you?" But even here the question is ritualistic because no one expects a truly thoughtful answer in reply to "How are you?" and, to the best of my knowledge, "Hello" is devoid of all meaning. These are just the small sounds that keep a community running smoothly, a few of the rituals of our society.

Much of the communication among lower animals is carried on in ritualistic fashion. Indeed, more of the communication among human beings is of a ritualistic nature than we in our thoughtful moments might care to admit. In this essay I shall describe the origins of some ritualistic behavior in lower forms. Eventually, however, we shall look at matters within human populations where our use of "ritual" may become somewhat distorted. Communication, according to a statement made earlier in this section, serves to keep the elements of a complex interacting system informed as to the overall state of affairs within the system. In human societies, complex systems themselves often serve as elements in an interacting system of a still higher order of complexity. The stance taken by a spokesman of a particular elemental system in his effort to preserve the larger complex is frequently adopted in order to impress his constituency; we shall refer to this stance as a ritualistic one. Of all ritualistic behavior in man, this is probably the most important, for it lies at the heart of power politics.

During the courtship of certain lovebirds the male lifts one foot in preparation for mating. Should the female repulse him at just that moment, he continues raising his foot and vigorously scratches his head. This is *displacement* scratching. A variety of similar displacement activities can be seen by watching persons unsuccessfully hailing taxicabs; the outstretched hand is seldom left idle once the cab has passed — it adjusts a hat, scratches a head, rubs imaginary dust from the eyes, or performs any of a number of other substitute actions. The teen-age boy who is rebuffed by his date when he attempts to put his arm around her in the movie is very much like the male lovebird — head scratching or hair grooming becomes a substitute excuse for having his upper arm projecting at shoulder height.

In certain species of lovebirds courtship behavior includes head-scratching by the male. What is a sign of disappointment (displacement scratching) in one species is an integral part of the courtship behavior in another and, in this context, it becomes part of the species' ritualistic behavior. The act has assumed an entirely different meaning as if a teen-age boy with a romantic interest in a young lady could signal this fact to her by passing his hand through his hair or by rubbing his chin. If I remember correctly, he can.

To an animal behaviorist, ritualistic behavior is behavior that has altered its original meaning and has adopted a new and perhaps quite different significance. In some duck species an aggressive female that is fighting with her neighbors from a positon of safety near her drake assumes a certain pose: her head is pointed to the rear so that she can watch the other pair of ducks and her chest faces the drake. Now, in other duck species, this particular pose has been incorporated into the courtship behavior of the female: head pointed back, chest first.

An obvious ritual for many men is that of cleaning, filling, tamping, and lighting their pipes during interviews or conferences. It is a time-consuming ritual that allows the performer to compose his thoughts. Furthermore, it is one that other persons seem loathe to interrupt, probably because pipe smoking and thoughtful contemplation are linked in many minds.

Much of the adverse reaction of the older generations (grandparents are nearly as numerous as parents) to today's youth stems from the contempt in which the young hold obvious rituals and ritualistic behavior. Clothes make the man? Obviously not in the eyes of many of our young persons. One hemline for all women? Never again, it seems. Bra-less girls according to one group of elder citizens are provocative. How's that? Certain manufacturing firms have been claiming the opposite for nearly as long as I can remember. Rituals, apparently, need not be consistent.

Among lower animals rituals keep simple groups operating smoothly. The young individual who is virtually defenseless before the onslaught of an irate adult saves himself by employing stereotyped reactions. A puppy, even a gangling half-grown puppy of a large breed of dog, can avoid harm by lying on its back; only rarely will an adult dog harm a puppy that assumes this position. Young birds avoid harm at the hands of their elders by making juvenile sounds and by assuming juvenile poses. The list can be extended to great length because in any species, but especially in those that have an elementary social structure, the interplay of territoritality, of individual needs, of mating behavior, and of rearing young can be surprisingly complex.

Modern man has built himself societies the likes of which have never existed before. Each empire that has arisen in the past has arisen around an effective (for its time) communication network. There is more than a grain of truth in the claim that all roads lead to Rome. Today the telephone and radio and TV networks carry messages to remote parts of the world instantaneously.

The complexity of modern society has rendered the town-meeting or city-state type of government impossible; political techniques suitable for a few thousand persons are not suitable for 200 million. Officially, the government of the United States is set up on the basis of geographically defined political units. Unofficially, there are large numbers of special interest groups, each exerting its influence on governmental processes. The country operates only to the extent that the spokesmen of these various groups successfully negotiate with one another while representing the short-term interests of their constituencies.

A spokesman on the political scene is effective only as long as he is retained by his constituency but, simultaneously, only if he is sufficiently astute to discern that what is best for the country as a whole need not appear to be best for his supporters. Conflicting demands on a politician can be met only by ritualistic stances. An expert on foreign affairs must appear indifferent to civil rights in order to remain in Washington. A spokesman for the black community must avoid any act that would lead to the label, "Uncle Tom." Years of experience are required to interpret the communiques issued by heads of state. The rituals that are demanded in public life serve an important function; they are dangerous when misread through inexperience or through design by purposeful men. As a society becomes increasingly complex, awareness of and allowances for the ritualistic behavior of others become essential for the very preservation of that society. Man lives in a ritualistically polyglot world. The need to preserve order both within and between complex subgroups severely taxes and may eventually overwhelm our communicative abilities.

On the Glib Assurance

"Salt is necessary for life."

Maple trees have been dying throughout the nation for a dozen years or more. The cause is not known; no virus or other disease organism has been identified as a likely culprit.

As long as causal organisms manage to elude detection, theories thrive. At one moment it appeared that the maples in towns were suffering more than those growing in woods or fields; furthermore, those at street intersections appeared to be in the worst condition of all. A pattern of dead and dying trees that follwed city streets with noticeable concentrations within natural drainage areas suggested that the salt used for snow removal rather than disease might be the killer. According to this theory, the use of salt on city streets for snow removal harmed the curbside trees by poisoning their root systems.

A federal report was issued that questioned the wisdom of using salt for snow removal. It mentioned damage to trees in Minnesota, citing the tendency for trees on boulevards to turn yellow or brown or, in some instances, to lose their leaves altogether.

The salt industry was fast to respond to this challenge. Salt, claimed a spokesman, is necessary for life — both plant and animal life. Some companies, he continued, use additives in their salt, but his firm avoided additives; it put out only pure rock salt.

The reassuring words about the necessity of salt for life were buttressed by claims that farmers buy salt to increase the yield of their crops and that the firm itself had evidence that salt stimulated plants to vigorous growth. Scarcely two years later the nation was horrified to learn just how poisonous an excessive amount of salt can be. Seven infants died after table salt was mistakenly substituted for sugar in making the babies' formula at an upstate New York hospital. Salt is indeed necessary for plant and animal life. The amount of salt that is required for life, however, falls within narrow limits. Too much salt need not be stimulating nor need it be harmless; indeed, too much salt (and that which goes on winter highways and streets is certainly not weighed out on delicate scales) can be deadly.

There are legitimate reasons for suspending judgment as to whether or not salt is responsible for the death of maple trees. Perhaps it has no role. Dead trees, for example, can be seen far from highways on remote mountain slopes. Perhaps salt has a subsidiary role in killing trees already damaged by disease, drought, or air pollutants. The role of salt is open to investigation; consequently, there are logical procedures by which appropriate studies can and should be carried out. The inexcusable aspect of the controversy over salt was the release to the public of a statement that exudes confidence but distorts or omits well-known facts. Here is what I call the "glib assurance": words uttered with no personal commitment on the part of the speaker, words designed more to mislead than to educate.

"CAUTION: Cigarette smoking may be hazardous. . . ."

The tobacco industry, in its effort to combat the Surgeon General's contention that smoking causes cancer, has stated its case in numerous full-page newspaper advertisements. Three of the points made in one of these advertisements illustrate once more what I call the glib assurance, an assurance given with no expectation that it will be challenged and with no intention of assuming moral responsibility should the stated claims prove to be wrong. These points and my response to each one are as follows:

Point 1. The genetic makeup of the individual largely determines his susceptibility to cancer. . . .
Response: True, but irrelevant. At no time has the Surgeon General suggested that smoking is the sole cause of cancer, nor has he suggested that all persons react to smoking in a uniform way. The federal government, to cite an analogous example, has established stringent regulations that limit the use of radioactive materials and radiation emitters; these regulations have been established because radiation is dangerous. That the precise extent of the danger depends in part upon the genetic makeup of the exposed individual has no bearing on the need for these safety regulations. Nor, in the case of automobile safety, does the genetic basis of an individual's height determine whether or not headrests should be required to minimize neck injuries.
Point 2. Many factors other than smoking are significantly associated with cancer. . . .
Response: This is a specious statement that has no bearing whatsoever on the Surgeon General's claim; its intended purpose is to focus our attention elsewhere.
Point 3. Statistical associations between smoking and lung cancer are not proof of causal relationship in the opinion of most epidemiologists.
Response: This statement represents a doubleheaded attack on the Surgeon General. At first glance it suggests that the Surgeon General has based his actions on the recommendations of a misinformed handful of medical advisors. A more careful reading shows that this interpretation may not be

correct, for the statement can also be interpreted as saying that *statistical associations are not proof*. The statement, if read in this way, does *not* say that most epidemiologists disagree with the conclusions or recommendations of the Surgeon General. All "scientific" conclusions are based on statistical associations; "proof" consists of a high correlation that leads to accurate predictive ability. No intelligent person interprets correlation alone as a cause-and-effect relationship.

Scientific papers, as we shall note elsewhere, tend to make dull reading. To a large extent the author's extreme caution is responsible. Seldom does the author of a technical paper omit details in order to lighten the task of the reader. If an important detail is omitted by accident, most scientists will supply the missing information promptly upon request. The advertising world operates quite differently. An advertisement is designed for instant consumption; it is designed to elicit instinctive or emotional rather than rational responses. Advertisements are not published in order to initiate a discussion. Their authors are rarely called upon to furnish background information; few would feel obliged to do so even if requested.

"The fun is long gone. . . .
The Game rolls on. . . .
The filth piles up."

During February, 1969, an oil well off the coast of Santa Barbara, California, blew during drilling operations. A fissure opened in the ocean floor. Oil and gas poured from the bottom of the sea to the surface where it boiled out at the rate of 1000 gallons per hour. The massive leak continued for 11 days or more. The oil slick grew until it covered more than 400 square miles of ocean and had fouled 40 miles of excellent southern California beaches.

The following selections from the New York Times* reveal the framework of business and governmental communication within which this oil spill occurred.

● ● ●

September 25, 1967

A small per cent of the Santa Barbara population is opposed to the sale.

T.D. Barrow
Senior Vice-President
Humble Oil and Refining Company

● ● ●

*© 1969 by The New York Times Company. Reprinted by permission.

December 12, 1967

We appreciate that there has been some opposition to the drilling; however, the reasons for the opposition seem indeed trivial as compared to the overall benefits to be desired.

>   Dixon Moorehead
>   Farmers National Bank of Santa Barbara

● ● ●

December 21, 1967

[The regional oil and gas supervisor is] vested with ample authority to take all steps economically and technologically feasible to minimize the number of structures and to take such other steps as are necessary to prevent pollution and minimize ecological and esthetic damage.

>   Harry R. Anderson
>   Assistant Secretary of the Interior

● ● ●

December 21, 1967

We feel maximum provision has been made for the local environment and that further delay in the lease sale would not be consistent with the national interest or regional economic welfare.

>   David S. Black
>   Undersecretary of the
>   Interior

● ● ●

February 17, 1968

It is essential that the Geological Survey, whose responsibility it is, take every precaution possible in its supervisory functions.
Pollution from oil spills in drilling production is an additional hazard which we must take every means to avoid.

>   Steward L. Udall
>   Secretary of the Interior

● ● ●

April 8, 1968

The heat has not died down but we can keep trying to alleviate the fears of the people in Santa Barbara.

>   J. Cordell Moore
>   Assistant Secretary of the Interior
>   for Mineral Resources

● ● ●

February 14, 1969

Mr. Chairman, I would like to comment further here: I think we have to look at these problems relatively. I am always tremendously impressed at the publicity that death of birds receives versus the loss of people in our country in this day and age. When I think of the folks that gave up their lives when they came down into the ocean off Los Angeles some three weeks ago and the fact that our society forgets about that within a 24-hour period, I think that relative to that the fact that we have had no loss of life from this incident is important.

> Fred C. Hartley
> President, Union Oil Company
> of California

• • •

March 8, 1969

The Federal Government has failed to maintain its share of responsibility for combatting pollution. . . . The oil companies have been left to their own devices, because no Federal funds were made available.

> Representative Bob Wilson
> Republican, California

• • •

June 29, 1969

Offshore mobile drilling rigs have multiplied to approximately 200 units worldwide in 1968. The number in operation has almost doubled in the past three years.

These complex operations are conducted with great care to assure proper environmental control. The responsible members of industry recognize their obligation to conduct these essential operations in a manner that will not destroy or damage the surrounding marine life and recreational areas.

U.S. offshore lands to a water depth of 600 feet, include 875,000 square miles and 1,370,000 square miles if this is extended to a depth of 6,000 feet. Only 19,000 square miles, or slightly more than 1 per cent, have been leased to date.

If the present trend of increased governmental control continues, the industry will never recover its investment and the ingenious off-shore industry will die.

> T.D. Barrow
> Senior Vice-President
> Humble Oil and Refining Company

On the Dullness
of Scientific Writing

John Steinbeck's French poodle, Charley, developed urinary problems as the two motored through northern Idaho. Several moments of the reader's time are taken to consider this minor misfortune:*

> I was having to make many stops for Charley's sake. Charley was having increasing difficulty in evacuating his bladder, which is Nellie talk for the sad symptoms of not being able to pee. This sometimes caused him pain and always caused him embarrassment. Consider this dog of great *élan*, of impeccable manner, of *ton, enfin* of a certain majesty. Not only did he hurt, but his feelings were hurt. I would stop beside the road and let him wander, and turn my back on him in kindness. It took him a very long time. If it happened to a human male I would have thought it was prostatitis. Charley is an elderly gentleman of the French persuasion. The only two ailments the French will admit to are that and a bad liver.

Here is a trivial event in Steinbeck's long life but an event duly recorded and appreciated. It would, after all, have been a matter of no import if Charley had died, then or earlier. But compare the pleasure gained by this small diversion to the agony of learning about human behavior.

> The importance of some of these primary reinforcers has been recognized only in recent years. In many cases, we still have much to discover about the operations that determine their reinforcing strengths. In other cases, such as exploration, novelty, and affection, the actual specification of the class of reinforcing stimuli is far from established. Though these problems present real difficulties for the development of drive concepts.**...

In this case we have a discussion of matters that touch on the behavior of each person throughout his life. It describes phenomena that can be used to manipulate voting patterns, shopping preferences, and work habits. These are not trivial matters. A president elected through the efforts of Madison Avenue sets the foreign policy of the United States; to a large extent his behavior determines whether the world is at peace or war. Corporations made large through the manipulation of buying habits tend to become still larger despite all efforts to control them. Indeed, the efforts at control are, in turn, subject to

---

*From *Travels with Charley in Search of America* by John Steinbeck. Copyright © 1961, 1962 by The Curtis Publishing Company, Inc. Copyright © 1962 by John Steinbeck. Reprinted by permission of The Viking Press, Inc.

**Reprinted from J. R. Millensen's, *Principles of Behavioral Analysis* by permission of the publisher, The Macmillan Company. Copyright, J. R. Millensen, 1967.

manipulation. Nevertheless, the paragraph that deals with such important matters turns most of us off. Such is the sad fate of scientific writing. But need it be so?

Why should scientists as an all-too-general rule write so poorly? First, they are not trained to write. Second, the importance of their message is in its content rather than its form. Third, the words available to them are very often defined nearly as rigorously as are mathematical symbols; in a sense, the hands of technical writers are tied.

Practice makes perfect; one learns to write by writing. A professional writer spends his life playing with words, learning shades of meaning, noting interesting, pleasant, or shocking combinations of sounds or thoughts. The scientist lacks the time for such needed practice. He manipulates symbols because they lend themselves to manipulation. Playing with words does not lead to problem solving.

The Old Testament lists certain dietary laws. Some of these are of a conditional nature:

> And every beast that parteth the hoof, and cleaveth the cleft into two claws, *and* cheweth the cud among the beasts, that ye shall eat. Nevertheless these ye shall not eat, of them that chew the cud, or of them that divide the cloven hoof; *as* the camel, and the hare, and the coney: for they chew the cud, but divide not the hoof; *therefore* they *are* unclean unto you. And the swine, because it divideth the hoof, yet cheweth not the cud, it *is* unclean unto you: ye shall not eat of their flesh, nor touch their dead carcass.

Understandable? Yes, but reducible to the general form $(x|y)$ by a mathematical mind accustomed to the use of mathematical symbols.

The human geneticist reduces pedigrees to charts in which circles and squares represent women and men; connecting lines represent marriages or offspring. Such charts can illustrate simple family relationships, complex family pedigrees such as those of the early Mormons where one man might have several wives and dozens of children, or the complex and intricate marriage customs of primitive peoples. To many persons the following epitaph may appear simpler than a diagram, but how many can quickly unravel the relationships described?

> Here lies the son; here lies the mother;
> Here lies the daughter; here lies the father;
> Here lies the sister; here lies the brother;
> Here lies the wife and the husband.
> Still, there are only three people here.

Not only is the scientist unskilled at the art of manipulating words, he is constrained by the use to which he would put those that he does use. Consider the outcome of an interview between a run-of-the-mill newspaper reporter and a scientist. The latter weighs each word he utters, constantly testing his statements to see if anything is being said that is misleading or that might be misconstrued by the eventual reader. Such a cautious and painstaking interview must be deadly for the reporter. Most of his time is spent listening to this professor-type backing and filling, constantly revising phrases by means of nearly imperceptible alterations. But imagine the shock the scientist gets when he reads the story as it finally appears in the local paper. Too often a reporter seems only to ask of himself "Is there a grain of truth in what I have written?" If so, he feels that he has fulfilled his journalistic obligation. Here is a contradiction in personal goals — the scientist's pathological fear of uttering a grain of untruth, and the reporter's joyous satisfaction at having incorporated even one grain of truth. It is a contradiction so difficult to resolve that good scientific reporters, those that do in fact resolve it satisfactorily, can be counted on the fingers of one's hands.

In his novel, *Madame Bovary*, Gustave Flaubert describes the onset of gangrene. He is not interested in gangrene as a disease. Nor is he interested in the clubfooted stable boy who eventually lost his leg because of the infection. Flaubert had one purpose in mind when he described the whole unpleasant episode — to point up the cupidity and stupidity of a character named Charles Bovary. It served his purpose to give an accurate description of the boy's suffering and of his rotting flesh because an accurate description of these things best served his other, more important purpose.

Flaubert's account would scarcely substitute for a clinical report intended to serve as instructional material in the training of nurses or young interns. For this we would need an account of the patient's temperature, a record of intake and outflow of fluids, comments on dehydration and shock, pulse, and numberless other facts and figures. The account of gangrene in the novel makes good (and, within limits, accurate) reading because extraneous, distracting details have been stripped away. Here lies another cause of the dullness of scientific prose — the need to record more than the average reader cares to learn. The more poorly the problem is understood, the greater the need to list details. How fortunate mathematicians are that $2 + 2 = 4$, whether the symbols are written on fine bond or yellow scratch paper, or whether they are written at all.

Words used in science are not entirely comparable to those used in everyday speech. The latter evolve rather rapidly; the former are supposed to remain fixed in agreement with recognized definitions. Salmonella is a bacterium that causes a variety of intestinal disturbances — severe and mild — in man. A Pacific Coast congressman, noting the similarity between the words "Salmonella" and "salmon" introduced a bill that would have changed the former. It was a ludicrous affair. Scientific names are neither invented nor destroyed by legislative acts. Scientific names are nearly sacred; furthermore, a great deal of minor scientific terminology is also sacrosanct. Even within a single technical paper, a given word may be defined for purposes of that paper alone so that whenever that word is used it means precisely what it is said to mean. "When *I* use a word," Humpty Dumpty said in a scornful voice, "it means just what I choose it to mean – neither more nor less." "The question is," he continued later, "which is to be master – that's all." He would have made a good writer of scientific reports.

A great deal of technical writing is duller than it need be. Effort is required to write well. Many scientists either do not have the stamina to ponder alternative ways of expressing a thought or are not convinced that the result is worth the investment.

Fast writing by a student is known as "exam-ese"; and by the administrator, "directive-ese." Neither is precise; neither says exactly what the writer intends. In effect, the student relies on the good will of the professor to read the correct answer into the words that he has hurriedly scribbled down. Similarly, the administrator trusts that his directive will fall into sympathetic hands and that recipients will act according to the intent rather than the actual words of the mineographed message. To an extent far greater than we would care to admit, scientific papers also rely upon a conscientious interpretation by their readers.

Too often too much is crowded into a scientific paper. The discussion of any one point is frequently dangerously attenuated while the author, like a child on an Easter egg hunt, relates his observations to all conceivable natural phenomena. If he only realized the number of readers who refuse to become involved! Many a discussion goes unread because it is not succinct enough for the reader's temper. One, two, or three well-made points are read and understood; five or six points are five or six points down the drain.

With this thought, this digression from biology ends.

The Flow of Information

Ours is an age of information explosion. Never before have so many worked so hard to discover new facts or reveal unseen relationships between old ones. Among the sciences, first physics had its renaissance; then biology (with the help of physical scientists) entered the era of molecular biology — an era that has seen the cracking of the genetic code and the unraveling of the feedback mechanisms that control gene action during development.

The explosion in the amount of information available for dissemination has overloaded and cracked traditional lines of communication. What are some of these lines? To what stresses have they been exposed? What are the consequences of irregular or faulty information flow? These are the sorts of questions we shall discuss in this brief essay.

The traditional mechanism for transmitting information from one generation to the next is the teacher. Today's teacher, like today's automobile, faces almost instant obsolescence.* Despite their research efforts, most teachers are reservoirs rather than fountainheads of knowledge. There was a time when the reservoir was filled during graduate-student days, and a lifetime of reasonably competent teaching was thereby assured. That time no longer exists. References to scientific papers published in 1950, references that appear in other articles within related fields, decrease with time in a manner such that their number after 6 years has dropped by half. References to papers published in 1955 drop in number more rapidly so that the "half-life" is only 3 years. References to papers published in 1960 have a half-life of only 1 1/2 years.

The amount of information a graduate student in the sciences must absorb is tremendous. Publications read at the onset of graduate training are largely out of date by the time the student schedules his final examination. The problem for the teacher is at least as severe, perhaps more so. A teaching career may extend for 30 years. If the half-life of relevant information is as short as 1 1/2 years, the teacher must prepare to revamp his knowledge every 3rd or 4th year in order to keep pace with his students and young colleagues. The interval between traditional sabbatical leaves is about five times the half-life of scientific information. A year's study every 7th year no longer serves to refurbish the teacher who, because of committee work or other administrative chores, has become so utterly out of date.

The flow of information from generation to generation is only one of the many paths that information must follow in a modern society. Lines of communication must be kept open between the various segments of society if

*Information for this section has been drawn in large part from D. L. Nanney, "Some Issues in Biology Teaching," *Bioscience*, 18 (1968), 104-107.

these are to function smoothly. Unfortunately, these lines are not always maintained. Too frequently, in fact, they are deliberately closed. How often do we learn that high officials, in the United States as well as elsewhere, surround themselves with men who "tell them what they want to hear"? Information is not what flows under such circumstances.

An analysis of the flow of information in respect to the control of ragweed has been made by Dr. F.E. Egler of Yale University. This study is described here because of the patterns of communication it has revealed.

The biological facts that are necessary in order to understand ragweed and the need for its control are relatively simple. Ragweed pollen is one of the chief offenders in causing hay fever. The ragweed is a "pioneer" plant that is capable of exploiting barren soil but is unable to compete in stable plant communities. Like many other plant species, it can be killed by modern chemical herbicides.

Control of ragweed by herbicidal sprays would appear to represent a frontal attack on the hay fever problem. Recommendations to this effect emerge from joint conferences of medical doctors and representatives of chemical industries. (see the illustration on page 221) The recommendations are transmitted to spray contractors and business associations. The industrial chemists, of course, keep both the sprayers and the conference members and associations informed about new developments in pesticide technology. On the chart this flow of information takes place along lines of communication that encircle "Industry" like a pinwheel.

The funds for ragweed control are largely community or governmental funds. Consequently, lines of communication are kept open between segments of industry (the source of herbicides) and government (the sources of funds). Industry also keeps in touch with citizens' groups since concerned citizens are those most likely to influence town boards and city councils and to request local ragweed control measures.

Groups of citizens who are concerned about hay fever and the eradication of ragweed frequently seek advice from state and federal agencies. The information they receive is, of course, very much like that supplied to them by industry, for the government itself relies heavily upon information that comes from industrial sources. The pinwheel is the source of virtually all relevant information.

Ragweed, as we mentioned above, is a pioneer weed that is adept at invading barren or disturbed soil. The application of herbicides over large areas converts these into large *barren* areas. The barren areas, in turn, are promptly invaded by ragweed; large stands of ragweed appear where previously there were mixed plant communities. In this manner, the application of herbicides exaggerates both the ragweed problem and its concomitant problem, hay fever.

For the most part, the ecologists in university biology departments know the natural history of ragweed very well. They do not take part in the conferences on hay fever however. These conferences are generally convened by

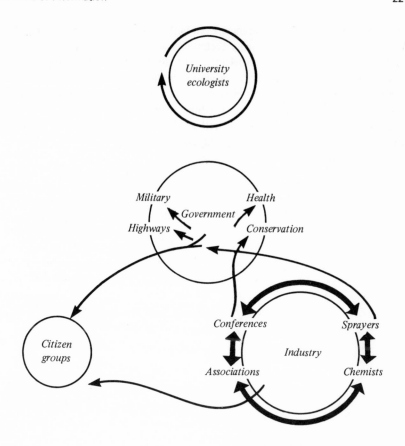

The flow of information concerning ragweed control. (*Adapted from* F.E. Egler, "Pesticides – in Our Ecosystem," *American Scientist,* 52 (1964), 110-136. *Reprinted* in *The Subversive Science* (Shepard and McKinley, Editors).

industry, perhaps at the instigation of medical doctors, to deal with medical problems. And so, for practical purposes, university ecologists have a separate line of communication that is independent of the others on the chart and feeds no information into the main system. As a result, the amount of herbicides used for hay fever control steadily increases as each application lays bare more ground for invasion by still more ragweed plants.

Faulty channels of communication lead to an improper distribution of information among those responsible for reaching decisions. In the short run, it may be pleasant to hear only what one wants to hear. This is true of presidents; it is true as well of herbicide manufacturers and other business executives. In the long run, however, it is disastrous to learn only part of any story and, therefore, to act in large part from ignorance. Even with all available evidence at hand, the long-range consequences of many technological feats are extremely difficult to foretell – the effects of a hydroelectric dam on the economy of a river basin, or of an intensive insect eradication program on the natural ecology of a country, or of an underground H-bomb blast on the geologic phenomena in the test area. Distortions in the types of information made available during preliminary discussions, willful distortions as well as those arising by neglect, render the arrival at a sound decision virtually impossible.

SECTION FOUR

Behavior

Introduction

Behavior is what man is all about. Human anatomy, human physiology, human genetics, and a host of other man-centered disciplines differ in only trivial ways from the corresponding branches of general vertebrate zoology. With behavior, however, we enter a realm in which man differs by several orders of magnitude from his nearest neighbors. The mind of man, not merely in the restricted sense of intelligence (whatever that is) but in the broad sense that includes human values and aspirations of all kinds, must form the "heart" of a study of man.

In the preceding sections, I have commented on the intertwining of sex and communication. Sandburg's characters were to a considerable extent communicating with one another, but the selections from *Always the Young Strangers* served to introduce sex. Moravia's and Michener's characters were engaged largely in sexual matters; nevertheless, they served to illustrate certain aspects of communication. And so it is with behavior. The mind and behavior are so interlaced that a section on behavior must of necessity include material on the central nervous system (especially the brain), the mind, and the manipulation of mental processes. A discussion of drugs must fall within the confines of human behavior; to avoid a discussion of drugs would, I believe, render this section vulnerable to devastating criticism. Similarly, to skirt around the question of public opinion, especially the molding of public opinion, would imply that I prefer "safe" topics to those that influence the direction in which society chooses to move.

Four selections have been chosen to introduce the matter of behavior. The first is a whimsical choice — an essay by Montaigne on the inconsistent behavior of man. Montaigne was the one who, in the Section on Genetics, puzzled over the inheritance of kidney stones. How, he asked, can a young healthy man transmit to his son an ailment from which the father himself first suffers only decades after his son's birth. Montaigne's questions about inheritance were good questions. He now returns with numerous observations on the seemingly unpredictable manner in which men (and women) act on successive days. The behavior of man is indeed erratic. The behavioral psychologist strives to understand this vacillation, and if he is either a politician, a military leader, or a salesman as well, he tries to influence the nature and direction of men's actions.

The second selection is an article by J.B.S. Haldane, one of the last articles written by one of the great biologists of this century. The cancer "near the hind end" of his intestine, as he describes it in his opening paragraph, proved to be fatal – a possibility foreseen by Haldane. The article summarizes Haldane's view of life and death and thus serves as a partial explanation, at least, for motivations that lay behind his own, often bizarre, actions. I also chose Haldane's article because, like many British scientists of his generation, he had experimented personally with a variety of drugs. Consequently, a message of great importance is hidden in his assertion that the "alterations of my consciousness due to . . . drugs were trivial compared with those produced in the course of my work." A lasting and worthwhile expansion of the mind comes from its use, not abuse. Among Haldane's last scientific contributions were solutions to mathematical problems arrived at in the hospital immediately following surgery for cancer. In a newspaper article prepared at this same time, Haldane points out that he made more original contributions to man's knowledge following his operation than most persons make in a lifetime; he took immense pride in intellectual accomplishments.

The third selection has been taken from *Growing Up Absurd* by Paul Goodman. Its choice coincided with a discussion I had had with a student on the possible existence of civilizations within which natural selection would favor low, rather than high, IQ's. During the discussion I denied that any civilization would exert such an influence. Later, perhaps under a continuing influence of McDermott's essay on technology, I recalled that modern society with its underuse of industrial employees and its increasing involvement with the appearance (for example, the corporate *image*) rather than the substance of things might be the very society that places a premium on intellectual mediocrity. Paul Goodman has developed a similar thesis in discussing the problems of modern youth; I have reproduced his chapter entitled "Jobs."

Disenchanted youngsters in large numbers have taken to drugs in attempting to escape what appears to be a mad world. This escape route itself is an insane one, particularly in the case of those drugs that lead to physiological dependence or true addiction. The last introductory selection is the summarizing lecture delivered at a British symposium on adolescent drug dependence. Throughout this text my aim has been to avoid the use of professional scientific literature as a source of introductory selections. In the case of Roy Goulding's summarizing address, however, I have made an exception. First, its presentation is almost conversational in form and so it makes for relaxed rather than hectic reading. Second, in touching on a wide variety of aspects of drug use, the summary reveals the thoughts of earlier speakers together with Dr. Goulding's reactions to them – almost in dialogue form. I have taken the liberty of removing Goulding's references to individual symposium participants because the summary reads more smoothly without the distracting references to persons otherwise unidentified.

Of the Inconsistency
of our Actions*

Montaigne

Those who make a practice of comparing human actions are never so perplexed as when they try to see them as a whole and in the same light; for they commonly contradict each other so strangely that it seems impossible that they have come from the same shop. One moment young Marius is a son of Mars, another moment a son of Venus. Pope Boniface VIII, they say, entered office like a fox, behaved in it like a lion, and died like a dog. And who would believe that it was Nero, that living image of cruelty, who said, when they brought him in customary fashion the sentence of a condemned criminal to sign: "Would to God I had never learned to write!" So much his heart was wrung at condemning a man to death!

Everything is so full of such examples — each man, in fact, can supply himself with so many — that I find it strange to see intelligent men sometimes going to great pains to match these pieces; seeing that irresolution seems to me the most common and apparent defect of our nature, as witness that famous line of Publilius, the farce writer:

> Bad is the plan that never can be changed.
>
> PUBLILIUS SYRUS

There is some justification for basing a judgment of a man on the most ordinary acts of his life; but in view of the natural instability of our conduct and opinions, it has often seemed to me that even good authors are wrong to insist on fashioning a consistent and solid fabric out of us. They choose one general characteristic, and go and arrange and interpret all a man's actions to fit their picture; and if they cannot twist them enough, they go and set them down to dissimulation. Augustus has escaped them; for there is in this man throughout the course of his life such an obvious, abrupt, and continual variety of actions that even the boldest judges have had to let him go, intact and unsolved. Nothing

*Reprinted from The Complete Essays of Montaigne, translated by Donald M. Frame, with the permission of the publishers, Stanford University Press. Copyright © 1948, 1957, & 1958 by the Board of Trustees of the Lelend Stanford Junior University.

is harder for me than to believe in men's consistency, nothing easier than to believe in their inconsistency. He who would judge them in detail and distinctly, bit by bit, would more often hit upon the truth.

In all antiquity it is hard to pick out a dozen men who set their lives to a certain and constant course, which is the principal goal of wisdom. For, to comprise all wisdom in a word, says an ancient [Seneca] , and to embrace all the rules of our life in one, it is "always to will the same things, and always to oppose the same things." I would not deign, he says, to add "provided the will is just"; for if it is not just, it cannot always be whole.

In truth, I once learned that vice is only unruliness and lack of moderation, and that consequently consistency cannot be attributed to it. It is a maxim of Demosthenes, they say, that the beginning of all virtue is consultation and deliberation; and the end and perfection, consistency. If it were by reasoning that we settled on a particular course of action, we would choose the fairest course — but no one has thought of that:

> He spurns the thing he sought, and seeks anew
> What he just spurned; he seethes, his life's askew.
>
> HORACE

Our ordinary practice is to follow the inclinations of our appetite, to the left, to the right, uphill and down, as the wind of circumstance carries us. We think of what we want only at the moment we want it, and we change like that animal which takes the color of the place you set it on. What we have just now planned, we presently change, and presently again we retrace our steps: nothing but oscillation and inconsistency:

> Like puppets we are moved by outside strings.
>
> HORACE

We do not go; we are carried away, like floating objects, now gently, now violently, according as the water is angry or calm:

> Do we not see all humans unaware
> Of what they want, and always searching everywhere,
> And changing place, as if to drop the load they bear?
>
> LUCRETIUS

Every day a new fancy, and our humors shift with the shifts in the weather:

> Such are the minds of men, as is the fertile light
> That Father Jove himself sends down to make earth bright.
>
> HOMER

We float between different states of mind; we wish nothing freely, nothing absolutely, nothing constantly. If any man could prescribe and establish definite laws and a definite organization in his head, we should see shining throughout his life an evenness of habits, an order, and an infallible relation between his principles and his practice.

Empedocles noticed this inconsistency in the Agrigentines, that they abandoned themselves to pleasures as if they were to die on the morrow, and built as if they were never to die.

This man[1] would be easy to understand, as is shown by the example of the younger Cato: he who has touched one chord of him has touched all; he is a harmony of perfectly concordant sounds, which cannot conflict. With us, it is the opposite: for so many actions, we need so many individual judgments. The surest thing, in my opinion, would be to trace our actions to the neighboring circumstances, without getting into any further research and without drawing from them any other conclusions.

During the disorders of our poor country,[2] I was told that a girl, living near where I then was, had thrown herself out of a high window to avoid the violence of a knavish soldier quartered in her house. Not killed by the fall, she reasserted her purpose by trying to cut her throat with a knife. From this she was prevented, but only after wounding herself gravely. She herself confessed that the soldier had as yet pressed her only with requests, solicitations, and gifts; but she had been afraid, she said, that he would finally resort to force. And all this with such words, such expressions, not to mention the blood that testified to her virtue, as would have become another Lucrece. Now, I learned that as a matter of fact, both before and since, she was a wench not so hard to come to terms with. As the story says: Handsome and gentlemanly as you may be, when you have had no luck, do not promptly conclude that your mistress is inviolably chaste; for all you know, the mule driver may get his will with her.

Antigonus, having taken a liking to one of his soldiers for his virtue and valor, ordered his physicians to treat the man for a persistent internal malady that had long tormented him. After his cure, his master noticed that he was going about his business much less warmly, and asked him what had changed him so and made him such a coward. "You yourself, Sire," he answered, "by delivering me from the ills that made my life indifferent to me." A soldier of Lucullus who had been robbed of everything by the enemy made a bold attack on them to get revenge. When he had retrieved his loss, Lucullus, having formed a good opinion of him, urged him to some dangerous exploit with all the fine expostulations he could think of,

> With words that might have stirred a coward's heart.
>
> HORACE

"Urge some poor soldier who has been robbed to do it," he replied;

---

[1] The disciplined man in the sentence before last.
[2] The religious civil wars between Catholics and Protestants, which lasted intermittently from 1562 to 1594.

> Though but a rustic lout,
> "That man will go who's lost his money," he called out;
>   HORACE

and resolutely refused to go.

We read that Sultan Mohammed outrageously berated Hassan, leader of his Janissaries, because he saw his troops giving way to the Hungarians and Hassan himself behaving like a coward in the fight. Hassan's only reply was to go and hurl himself furiously — alone, just as he was, arms in hand — into the first body of enemies that he met, by whom he was promptly swallowed up; this was perhaps not so much self-justification as a change of mood, nor so much his natural valor as fresh spite.

That man whom you saw so adventurous yesterday, do not think it strange to find him just as cowardly today: either anger, or necessity, or company, or wine, or the sound of a trumpet, had put his heart in his belly. His was a courage formed not by reason, but by one of these circumstances; it is no wonder if he has now been made different by other, contrary circumstances.

These supple variations and contradictions that are seen in us have made some imagine that we have two souls, and others that two powers accompany us and drive us, each in its own way, one toward good, the other toward evil; for such sudden diversity cannot well be reconciled with a simple subject.

Not only does the wind of accident move me at will, but, besides, I am moved and disturbed as a result merely of my own unstable posture; and anyone who observes carefully can hardly find himself twice in the same state. I give my soul now one face, now another, according to which direction I turn it. If I speak of myself in different ways, that is because I look at myself in different ways. All contradictions may be found in me by some twist and in some fashion. Bashful, insolent; chaste, lascivious; talkative, taciturn; tough, delicate; clever, stupid; surly, affable; lying, truthful; learned, ignorant; liberal, miserly, and prodigal: all this I see in myself to some extent according to how I turn; and whoever studies himself really attentively finds in himself, yes, even in his judgment, this gyration and discord. I have nothing to say about myself absolutely, simply, and solidly, without confusion and without mixture, or in one word. Distinguo is the most universal member of my logic.

Although I am always minded to say good of what is good, and inclined to interpret favorably anything that can be so interpreted, still it is true that the strangeness of our condition makes it happen that we are often driven to do good by vice itself — were it not that doing good is judged by intention alone.

Therefore one courageous deed must not be taken to prove a man valiant; a man who was really valiant would be so always and on all occasions. If valor were a habit of virtue, and not a sally, it would make a man equally resolute in any contingency, the same alone as in company, the same in single combat as in battle; for, whatever they say, there is not one valor for the pavement and another for the camp. As bravely would he bear an illness in his bed as a wound in camp, and he would fear death no more in his home than in an assault. We would not see the same man charging into the breach with brave assurance, and later tormenting himself, like a woman, over the loss of a lawsuit or a son. When, though a coward against infamy, he is firm against poverty; when, though weak against the surgeon's knives, he is steadfast against the enemy's swords, the action is praiseworthy, not the man.

Many Greeks, says Cicero, cannot look at the enemy, and are brave in sickness; the Cimbrians and Celtiberians, just the opposite; for nothing can be uniform that does not spring from a firm principle.

There is no more extreme valor of its kind than Alexander's; but it is only of one kind, and not complete and universal enough. Incomparable though it is, it still has its blemishes; which is why we see him worry so frantically when he conceives the slightest suspicion that his men are plotting against his life, and why he behaves in such matters with such violent and indiscriminate injustice and with a fear that subverts his natural reason. Also superstition, with which he was so strongly tainted, bears some stamp of pusillanimity. And the excessiveness of the penance he did for the murder of Clytus is also evidence of the unevenness of his temper.

Our actions are nothing but a patch — they despise pleasure, but are too cowardly in pain; they are indifferent to glory, but infamy breaks their spirit — and we want to gain honor under false colors. Virtue will not be followed except for her own sake; and if we sometimes borrow her mask for some other purpose, she promptly snatches it from our face. It is a strong and vivid dye, once the soul is steeped in it, and will not go without taking the fabric with it. That is why, to judge a man, we must follow his traces long and carefully. If he does not maintain consistency for its own sake, with a way of life that has been well considered and preconcerted; if changing circumstances make him change his pace (I mean his path, for his pace may be hastened or slowed), let him go: that man goes before the wind, as the motto of our Talbot says.

It is no wonder, says an ancient [Seneca], that chance has so much power over us, since we live by chance. A man who has not directed his life as a whole toward a definite goal cannot possibly set his particular actions in order. A man who does not have a picture of the whole in his head cannot possibly arrange the pieces. What good does it do a man to lay in a supply of paints if he does not know what he is to paint? No one makes a definite plan of his life; we think about it only piecemeal. The archer must first know what he is aiming at, and then set his hand, his bow, his string, his arrow, and his movements for that goal. Our plans go astray because they have no direction and no aim. No wind works for the man who has no port of destination.

I do not agree with the judgment given in favor of Sophocles, on the strength of seeing one of his tragedies, that it proved him competent to manage his domestic affairs, against the accusation of his son. Nor do I think that the conjecture of the Parians sent to reform the Milesians was sufficient ground for the conclusion they drew. Visiting the island, they noticed the best-cultivated land and the best-run country houses, and noted down the names of their owners. Then they assembled the citizens in the town and appointed these owners the new governors and magistrates, judging that they, who were careful of their private affairs, would be careful of those of the public.

We are all patchwork, and so shapeless and diverse in composition that each bit, each moment, plays its own game. And there is as much difference between us and others. Consider it a great thing to play the part of one single man [Seneca]. Ambition can teach men valor, and temperance, and liberality, and even justice. Greed can implant in the heart of a shop apprentice, brought up in obscurity and idleness, the confidence to cast himself far from hearth and home, in a frail boat at the mercy of the waves and angry Neptune; it also teaches discretion and wisdom. Venus herself supplies resolution and boldness to boys still subject to discipline and the rod, and arms the tender hearts of virgins who are still in their mothers' laps:

> Furtively passing sleeping guards, with Love as guide,
> Alone by night the girl comes to the young man's side.
>   TIBULLUS

In view of this, a sound intellect will refuse to judge men simply by their outward actions; we must probe the inside and discover what springs set men in motion. But since this is an arduous and hazardous undertaking, I wish fewer people would meddle with it.

On Being Finite*

J.B.S. Haldane

I have reached the age of seventy-one, and been operated for a cancer near the hind end of my intestine. These events lead me to attend rather more closely than I did twenty years ago to the fact that I shall die within a few years — perhaps one year if the cancer has sent a colony of cells to another part of my body, perhaps twenty-five if it has not, and the rest of my cells behave unusually well.

Death, as something unavoidable, is probably quite a recent human discovery. Primitive men saw plenty of others die, but to judge from the bones of palaeolithic men, they very rarely saw people over the age of sixty, still less did they see deaths which could only be ascribed to old age. Almost all deaths were accidental. In a very sparse population infectious disease was probably rarer than it was in my youth in England, or is in India today. Perhaps it was as rare as in modern Britain; and when it happened was attributed to black magic.

Birds are like that today. They have a heavy death rate as eggs and nestlings. But once they can fly, as almost all can at a year old, the death rate is independent of age. For example, about two-thirds of all adult robins will be dead a year hence. So only one in two thousand or so lives to be seven years old. But if it does, it is no more likely to die in the next year than a year-old robin. Perhaps one robin in a million dies of old age — that is to say because one of its organs is worn out. If so, we have as yet no idea at what age such deaths occur.

Some animal species, on the other hand, have a much more sharply limited life span than men. For example, may-flies live for a year, almost all under water. They die within a day of emergence from water into air. Many moths, including most whose caterpillars make silk, cannot eat or drink, and live for only a week or two after emerging from their cocoons. Probably, however, most animals, even if they die of old age when protected by men, die of 'accident', usually by being eaten, or of disease, in their normal environment.

To arrive at the notion 'I may die' requires a certain amount of social consciousness. It is one thing to see other humans die, and a different one to

*On Being Finite by J.B.S. Haldane appeared in The Rationalist Annual for the Year 1965. Reprinted by permission of the publisher, Pemberton Publishing Co. Ltd. © The Pemberton Publishing Co., 1965.

realize that one is also human, and may die; though it must have been fairly obvious, if a tiger or a falling rock killed one's neighbour, that this could happen to oneself. Probably this was understood, at least by some people, several million years ago, long before our ancestors were recognizably men. They were presumably frightened, as we still are, by large and fierce animals not confined behind bars or a ditch, by rapidly rising floods, and so on. But if they had any language it is unlikely that it included the abstract notion of death, though it may have included words for several methods of killing and dying.

I suspect that the notion 'I must die' was a very much later development. I doubt if it could have arisen without some oral tradition about the past of one's tribe. It must have been a distressing discovery made by many different people, but not a secure part of the human tradition until language had become both fairly precise and fairly abstract. It was extremely difficult to accept, because it is unimaginable. One can imagine oneself playing a golden harp in heaven, in a fire in hell, or reborn as a cockroach, but one cannot imagine nothing. Most humans now believe, and probably most have believed for many thousands of years, in some kind of life after death.

There seems to me to be very little evidence in favour of this belief. What there is comes from one of four sources. These are revelation, memory of past lives, interviews with ghosts, and metaphysics. Various men and a few women have so impressed their contemporaries that their statements about a future life have been widely believed. Neither Jesus nor St Paul, who are the main sources of Christian revelation, have left clear accounts of how they were informed as to the future life. Muhammad said that he repeated the words of Jibril (Gabriel). According to the Hindu epic, the Mahabharata, Krishna, a man who subsequently died, claimed explicitly to be the supreme god, and taught Arjuna accordingly. The revelations are very different. According to almost, but not quite, all Hindu teachers, and to Buddhists, Jains, and members of related religions, almost everyone will be born again; and even the wickedest, after considerable punishment, will have another chance. Christianity and Islam offer no such consolation. Unfortunately, if one of these two religions is true, adherents of the other will be tormented eternally in hell. Of course some modern Christians and Muslims try to ignore the passages in their sacred books where such torment is described and promised. If the books are true, they are themselves in danger of hell fire.

Many adherents of the Indian religions have claimed to remember former lives on earth. This doctrine has the unfortunate effect that when its believers see a man or woman suffering they may say that he or she must have been very wicked in a former life to deserve such pain. Of course, every religion or philosophy which teaches that the world is administered by a just god or in accordance with a just law of karma leads to a similar conclusion. Fortunately there have always been enough adherents of religion with decent human feelings to prevent the full consequences of this doctrine from being worked out and the unfortunate completely neglected.

I think that were I a believer in rebirth I could have come to think that I remembered at least incidents in some past lives. In December 1916 I arrived at a camp behind the British trench line in what is now Iraq, after a night journey up the Tigris. When I got up next morning and looked at the Persian mountains on the eastern skyline I got the feeling that they were absolutely familiar, as if I had seen them thousands of times already. This may have been a by-product of my satisfaction in rejoining old friends in the Black Watch after a voyage of two months round the Cape via Bombay. It may of course have been a memory of a past life. I doubt this; but had I been subjected to one of the Indian equivalents of hypnosis I might well have come to believe that I remembered details of a life in ancient Mesopotamia or medieval Iraq. When such alleged memories have led to the decipherment of even one previously unknown ancient language or the excavation of a previously unknown ancient site, I shall be prepared to consider them more seriously.

I do not doubt that many people have seen and heard ghosts. In some cases a number of people saw the same ghost. Similarly in 1917 a large number of people saw the sun moving about in the sky and descending towards them at Fatima in Portugal. Detailed accounts may be bought at any Catholic bookshop. In fact, however, the vast majority of the human race saw nothing. And the miracle may be described as a collective hallucination, though of course this phrase does not explain it. Some day, I hope, such facts will be explained. Ghosts are reported particularly in places where people have been murdered, or have died after doing actions of which they were grossly ashamed. The paucity of ghosts on the Western Front of the First World War raises difficulties for those who believe in their existence, for many of the millions who died there resented their deaths as bitterly as the victims of civil murder. Ghosts are said sometimes to have communicated facts such as the identity of their murderers and the location of hidden treasure; and of course ghosts at spiritualistic seances often purport to give quite lengthy communications about their past lives. But even if we accept these statements as true there is no evidence that ghosts are conscious. The dead can influence the living; for example, someone will probably read this article after my death, and may agree with some of it. But that does not prove that I shall still be conscious. I think the correct conclusion from the facts before us is to keep an open mind on hauntings, messages through mediums, and so on, but to agree with what, according to the Brihadaranyaka Upanisad, the sage Yajnavalkya said two thousand five hundred or more years ago: 'Na pretya samjh n' asti' ('There is no consciousness for a ghost').

The metaphysical argument is roughly that the human mind is of a quite different nature to material objects and, unlike them, cannot be destroyed. Bertrand Russell has pointed out that this argument, if correct, proves that the human mind or soul is not merely unbounded in time but in space. Not only have I existed for ever, and will exist for ever, but I am aware of events at any distance, however great, and can influence them. This is contrary to the facts. But some Indian philosophers have accepted it. They have been left to explain the nature of the illusion which leads me to believe that I am not an almighty and omnipresent being. I can only say that any metaphysical theories produced by, or believed by, beings subject to so vast an illusion are highly suspect.

I do not regard the statement that all men desire a longer, or even an eternal life, as an adequate argument. To begin with, it is untrue. The Indian religions are largely concerned with escape from rebirth. Even if it were true of most people, we all of us, at one time or another, desire impossible things in this present life. I have no personal desire to be born again. If I am to be replaced, I would prefer to be replaced by someone without some of my congenital deficiencies, for example tone-deafness.

I think, then, that we must, at least provisionally, accept the notion that we are finite in time as we are in space, and act on this acceptance. This means that we must be, to some extent, Epicureans, simply because Epicurus was the first man who did his best to work out the consequences of his finitude and act on them. Of course, nobody living today can accept all Epicurus's teaching and example, for two good reasons. A great deal more is known about external nature, and at least something more about human nature, than Epicurus knew. And human society is very different from that of his time, though some people think it is becoming more like that from which he did his best to withdraw. By the way, not all professed Epicureans withdrew from public affairs. The divine Julius Caesar and his principal murderer, Caius Cassius, both belonged to this sect.

Epicurus taught that we should not be afraid of our own death, which amounts to being afraid of nothing. It is, however, rational to be afraid of the deaths of our friends, which are events of which we are aware, as our own death is not. I doubt whether we have an instinctive fear of death: but we certainly seem to have untaught fears of some ways of dying, such as falling from heights, being crushed, being eaten by large animals, and various forms of human violence. These fears cannot perhaps be abolished, but they can be overcome. One of the least plausible statements which Shakespeare attributed to Julius Caesar is:

> Cowards die many times before their death,
> The valiant never taste of death but once.

I don't think death has a taste, but its accompaniments have, and one only becomes valiant by sipping this taste, which is bitter at first, but soon becomes pleasant. If it did not, very few people would indulge in rock climbing, motor cycle racing, and similar activities. The pioneers, and a few people who make new records today, do such things to gain fame, but most people do so because they enjoy it. I am told that they have an unconscious death wish. Why not, if they do not let it dominate them? They enjoy flirting with Death, whom I picture as a woman, without intending to go all the way, though prepared to do so when the occasion arises. I disagree equally with Swinburne's line

Some gainless glimpse of Proserpine's veiled head.

I think such glimpses are gainful. They develop the virtue of courage.

It is commonly stated by Christians that the spirit yearns for immortality. Epicurus (or whoever wrote 'The principal doctrines') thought otherwise. 'The body', he wrote, 'perceives the limits of pleasure as infinite, and infinite time is required to supply it. But the mind, after attaining a reasoned understanding of the ultimate good of the body and its limits, and after dissipating fears as to the future, supplies us with the complete life, and we have no more need of infinite time: but neither does the mind shun pleasure, nor, when circumstances begin to bring about the departure from life, does it approach its end as though it fell short in any way of the best life.'

Epicurus believed in the existence of gods. He thought that people sometimes saw them, though the world carried on without them under the influence of natural forces. I believe in them a little less than he did. I believe, with the author of the relevant passage in the Brihadaranyaka Upanisad, that they are social products, or with Lenin that they are sterile but genuine flowers on the tree of human culture. So I try to be polite to the deities of the country in which I am living, except in so far as their reported conduct, like that of Sitala the Indian smallpox goddess, who is an anti-vaccinationist, is grossly anti-social.

Epicurus taught that we should eat well, but in moderation. His teaching on sexual activity was similar. He taught that marriage and parentage should be avoided, but that friendship (the word could also be translated as mutual love) is by far the greatest good. His disciples, who included women, lived together in a garden in Athens. He believed that the good were happier than the bad, but the virtues which he stressed were prudence, justice, and temperance rather than the intellectual or the heroic virtues. He recommended complete withdrawal from public affairs, which, in his time, when public affairs were largely wars between absolute monarchs, was sound advice.

He and his disciples were probably happy because they were busy working out a set of rather over-simplified scientific principles which can be read in Lucretius's great poem. But only a few people in any generation have ever found happiness in this way. I doubt if happiness is possible unless one works fairly hard and enjoys one's work. Aristotle taught that happiness is good activity, and in India Krishna taught that we should work for the sake of the work, not of its reward. I am in full agreement. What precisions does the acceptance of finitude bring to this notion?

The most satisfying work for most people is probably that which brings obvious and immediate benefits to a number of other people whom the worker meets personally, and for which he or she does not have to extract money or gifts from them — for example, the work of a postman, or a doctor under the National Health Service. The latter is, however, still somewhat bedevilled by the fact that, on economic grounds, he generally takes on a few more patients than the number which he could treat most efficiently. So is that of a teacher by the fact that many of his or her pupils do not want to be taught. Scientific research and artistic and literary creation are, at best, the most satisfying of all pursuits. But scientific research is being more and more debased by team work, in which a large number of workers do what they are told to do, not what they want to do, and the results are remote. Similarly, literary and artistic work usually involves, and probably always has involved, either conformity to the tastes of the rich or powerful or a constant struggle against poverty. In the past one painted a dishonest merchant praying to his favourite saint, or wrote an ode to an unattractive but influential duchess. Nowadays one paints designs to promote the sale of food which one would not eat, or labels the villain of one's novel a communist or capitalist according to one's geographical location.

I am not arguing that the community should keep artists to do 'art for art's sake' or scientists to do 'science for science's sake'. I argue that it should be possible for artists and scientists whose work is not paid to earn their living in other ways, and have sufficient leisure and spare money for the job which seems to them to make life worth living. During my first three years at a British university I had between twenty and thirty hours' teaching and administration per week. I did enough research in my spare time to get a job with much less teaching.

Of course, I am told that one should devote one's life to a great cause. Many of my contemporaries devoted themselves to the maintenance, and even the expansion, of the British Empire. Except for a few troubled spots like Aden, the British Empire is dead, and its servants who survive it are disillusioned old men and women. Those who worked for the World Revolution, whether primarily against capitalism or colonialism, were farther sighted, and have been more fortunate, including, in my opinion, those who died fighting. But the world revolution, like a swift stream, has its eddies, including recrudescences of various kinds of tyranny.

The trouble about great causes is that they are apt, in practice, to be great quarrels, and human beings find quarrelling enjoyable. They also give scope for power addicts, that is to say people who enjoy ordering other people about, as opposed to influencing them by example or argument. I am not happy working under a power addict, however able, efficient, and even kindly he may be. Some people are; but a power addict is a very brittle idol compared with a great cause, which may at least outlast one's life.

There is no doubt, however, that in identifying oneself with a great cause one can achieve a partial escape from that aspect of finitude called death. One can do worse than read such a book as Gordon Childe's What Happened in

History to learn which human achievements have been relatively permanent. The most permanent have, I suppose, been advances in technology. Curiously enough, some of the ancient ones may outlast the modern. The inventors of the lamp, the loom, and the saddle would recognize the descendants of their prototypes, even if the inventor of the first lamp might find it hard to believe that the lamp in his room was lit by coal burned a hundred miles away, and the saddle of a bicycle might seem rather small. Modern inventions are shorter lived. I doubt if there will be a gramophone outside a museum a century hence. But if men survive there will be far more compact means for reproducing music and speech. Edison will not have lived in vain. If Marx was right, and I happen to agree with him, inventors are doing more for the world revolution than many politicians. Changes in productive 'forces' generate a social instability which brings about change in productive relations, and hence in social structure. Those inventors who are Marxists are quite aware of this.

I venture to think that some of my scientific discoveries will outlive me, not merely for a human lifetime, but for very much longer. My reason for this belief is curious. I get credit in scientific literature for the less important of them; for example, 'The ingestion of even a quarter mol of magnesium chloride causes a marked acidosis (Haldane 1925)'. But the more important — for example, that the rate of mutation of a human gene can be measured — are taken for granted. Few geneticists, and much fewer of the people who make public pronouncements on human mutation, know that I made this discovery, much less where I published it. No greater compliment can be paid to a scientist than to take his original ideas for granted as part of the accepted framework of science during his lifetime.

Literature and the arts are more chancy. Consider Cinna. Catullus, one of his two great contemporary Latin poets, though Catullus's poems were only preserved by a stroke of luck, wrote of him

Smyrnam incana diu saecula pervoluent.

(The hoary centuries will long unroll 'Smyrna'.) I doubt if a line of this work survives, and its author is remembered for the Fourth Citizen's maniacal exclamation 'Tear him for his bad verses', in Act III, Scene III, of Julius Caesar. In fact he is not all dead, for his verses almost certainly influenced later Roman poets and hence those of recent centuries throughout the world. On the other hand, Cinna's younger contemporary Horace was right when he said that he had made a monument more lasting than brass and higher than the Pyramids. So, we may safely conjecture, was Shakespeare when he wrote

Not marble, nor the gilded monuments
Of princes, shall outlive this powerful rhyme.

But I wonder how many of Shakespeare's forgotten contemporaries thought the same. In the past, sculpture and mosaics had some chance of surviving two thousand years, paintings a much smaller one, though the murals of a few palaeolithic and Indian caves have been preserved. Music beyond a few simple traditional tunes had none at all. Nowadays the opposite is true. Paintings deteriorate, and may be burned or slashed. Sculpture will not stand up to modern weapons. But music can be recorded both as heard and in symbols. So Stravinsky and Hindemith are perhaps more likely to be directly known two thousand years hence than Picasso and Epstein.

Whether such an attitude is rational or not, there is no reasonable doubt that the prospect of death is greatly alleviated by the beliefs that one's work will be of use or ornament after that event, and that others will carry on with it.

I should find the prospect of death annoying if I had not had a very full experience mainly stemming from my work. I missed many possibilities because I got a severe wound in my right arm in 1915. However, adapting to this wound has been an experience denied to most people. One thing which I am really sorry to have missed is walking to France on the sea bottom, which incidentally would have involved some interesting physiological research beforehand. I only got the money needed for this purpose at the age of seventy. Most of my joyful experiences have been by-products. Thus I have enjoyed the embraces of two notoriously beautiful women. In neither case was there any wooing. After knowing one another for some time we felt like that, and no words or gifts were needed. I later married a colleague. We had known one another for some years, and our love was largely based on respect for each other's work. I have tried morphine, heroin, and bhang and ganja (hemp prepared for eating and smoking). The alternations of my consciousness due to these drugs were trivial compared with those produced in the course of my work. I once dreamed that I was reading a life of Christ written and illustrated by Edward Lear. But I can only remember Pontius Pilate's moustache. If you want a dream as original as that, don't take opium, but eat sixty grams of hexahydrated strontium chloride. I have had some of the standard adventurous experiences such as being pulled out of a crevasse in a glacier, and more which are unusual. For example, I was one of the first two people to pass forty-eight hours in a miniature submarine, and one of the first few to get out of one under water. I doubt whether, given my psychological make-up, I should have found many greater thrills in a hundred lives. So when the angel with the darker drink at last shall find me by the river's brink, and offering his cup, invite my soul forth to my lips to quaff, I shall not shrink.

Jobs*

Paul Goodman

1.

It's hard to grow up when there isn't enough man's work. There is "nearly full employment" (with highly significant exceptions), but there get to be fewer jobs that are necessary or unquestionably useful; that require energy and draw on some of one's best capacities; and that can be done keeping one's honor and dignity. In explaining the widespread troubles of adolescents and young men, this simple objective factor is not much mentioned. Let us here insist on it.

By "man's work" I mean a very simple idea, so simple that it is clearer to ingenuous boys than to most adults. To produce necessary food and shelter is man's work. During most of economic history most men have done this drudging work, secure that it was justified and worthy of a man to do it, though often feeling that the social conditions under which they did it were not worthy of a man, thinking, "It's better to die than to live so hard" — but they worked on. When the environment is forbidding, as in the Swiss Alps or the Aran Islands, we regard such work with poetic awe. In emergencies it is heroic, as when the bakers of Paris maintained the supply of bread during the French Revolution, or the milkman did not miss a day's delivery when the bombs recently tore up London.

At present there is little such subsistence work. In Communitas my brother and I guess that one-tenth of our economy is devoted to it; it is more likely one-twentieth. Production of food is actively discouraged. Farmers are not wanted and the young men go elsewhere. (The farm population is now less than 15 per cent of the total population.) Building, on the contrary, is immensely needed. New York City needs 65,000 new units a year, and is getting, net, 16,000. One would think that ambitious boys would flock to this work. But here we find that building, too, is discouraged. In a great city, for the last twenty years hundreds of thousands have been ill housed, yet we do not see science, industry, and labor enthusiastically enlisted in finding the quick solution to a definite problem. The promoters are interested in long-term investments, the real estate men in speculation, the city planners in votes and graft. The building craftsmen cannily see to it that their own numbers remain few, their methods antiquated, and their rewards high. None of these people is much interested in providing shelter, and nobody is at all interested in providing new manly jobs.

Once we turn away from the absolutely necessary subsistence jobs, however, we find that an enormous proportion of our production is not even unquestionably useful. Everybody knows and also feels this, and there has recently been a flood of books about our surfeit of honey, our insolent chariots, the follies of exurban ranch houses, our hucksters and our synthetic demand. Many acute things are said about this useless production and advertising, but not much about the workmen producing it and their frame of mind; and nothing at all, so far as I have noticed, about the plight of a young fellow looking for a manly occupation. The eloquent critics of the American way of life have themselves been so seduced by it that they think only in terms of selling commodities and point out that the goods are valueless; but they fail to see that people are being wasted and their skills insulted. (To give an analogy, in the many gleeful onslaughts on the Popular Culture that have appeared in recent years, there has been little thought of the plight of the honest artist cut off from his audience and sometimes, in public arts such as theater and architecture, from his medium.)

What is strange about it? American society has tried so hard and so ably to defend the practice and theory of production for profit and not primarily for use that now it has succeeded in making its jobs and products profitable and useless.

2.

Consider a likely useful job. A youth who is alert and willing but not "verbally intelligent" — perhaps he has quit high school at the eleventh grade (the median), as soon as he legally could — chooses for auto mechanic. That's a good job, familiar to him, he often watched them as a kid. It's careful and dirty at the same time. In a small garage it's sociable; one can talk to the customers (girls). You please people in trouble by fixing their cars, and a man is proud to see rolling out on its own the car that limped in behind the tow truck. The pay is as good as the next fellow's, who is respected.

So our young man takes this first-rate job. But what when he then learns that the cars have a built-in obsolescence, that the manufacturers do not want them to be repaired or repairable? They have lobbied a law that requires them to provide spare parts for only five years (it used to be ten). Repairing the new cars is often a matter of cosmetics, not mechanics; and the repairs are pointlessly expensive — a tail fin might cost $150. The insurance rates therefore double and treble on old and new cars both. Gone are the days of keeping the jalopies in good shape, the artist-work of a proud mechanic. But everybody is paying for foolishness, for in fact the new models are only trivially superior; the whole thing is a sell.

It is hard for the young man now to maintain his feelings of justification, sociability, serviceability. It is not surprising if he quickly becomes cynical and time-serving, interested in a fast buck. And so, on the notorious Reader's Digest test, the investigators (coming in with a disconnected coil wire) found that 63 per cent of mechanics charged for repairs they didn't make, and lucky if they didn't also take out the new fuel pump and replace it with a used one (65 per cent of radio repair shops, but only 49 per cent of watch repairmen "lied, overcharged, or gave false diagnoses").

There is an hypothesis that an important predisposition to juvenile delinquency is the combination of low verbal intelligence with high manual intelligence, delinquency giving a way of self-expression where other avenues are blocked by lack of schooling. A lad so endowed might well apply himself to the useful trade of mechanic.

3.

Most manual jobs do not lend themselves so readily to knowing the facts and fraudulently taking advantage oneself. In factory jobs the workman is likely to be ignorant of what goes on, since he performs a small operation on a big machine that he does not understand. Even so, there is evidence that he has the same disbelief in the enterprise as a whole, with a resulting attitude of profound indifference.

Semiskilled factory operatives are the largest category of workmen. (I am leafing through the U. S. Department of Labor's Occupational Outlook Handbook, 1957.) Big companies have tried the devices of applied anthropology to enhance the loyalty of these men to the firm, but apparently the effort is hopeless, for it is found that a thumping majority of the men don't care about the job or the firm; they couldn't care less and you can't make them care more. But this is not because of wages, hours, or working conditions, or management. On the contrary, tests that show the men's indifference to the company show also their (unaware) admiration for the way the company has designed and manages the plant; it is their very model of style, efficiency, and correct behavior. (Robert Dubin, for the U. S. Public Health Service.) Maybe if the men understood more, they would admire less. The union and the grievance committee take care of wages, hours, and conditions; these are the things the workmen themselves fought for and won. (Something was missing in that victory, and we have inherited the failure as well as the success.) The conclusion must be that workmen are indifferent to the job because of its intrinsic nature:

it does not enlist worth-while capacities, it is not "interesting"; it is not his, he is not "in" on it; the product is not really useful. And indeed, research directly on the subject, by Frederick Herzberg on Motivation to Work, shows that it is defects in the intrinsic aspects of the job that make workmen "unhappy." A survey of the literature (in Herzberg's Job Attitudes) shows that Interest is second in importance only to Security, whereas Wages, Conditions, Socializing, Hours, Ease, and Benefits are far less important. But foremen, significantly enough, think that the most important thing to the workman is his wages. (The investigators do not seem to inquire about the usefulness of the job — as if a primary purpose of working at a job were not that it is good for something! My guess is that a large factor in "Security" is the resigned reaction to not being able to take into account whether the work of one's hands is useful for anything; for in a normal life situation, if what we do is useful, we feel secure about being needed. The other largest factor in "Security" is, I think, the sense of being needed for one's unique contribution, and this is measured in these tests by the primary importance the workers assign to being "in" on things and to "work done being appreciated." Table prepared by Labor Relations Institute of New York.)

Limited as they are, what a remarkable insight such studies give us, that men want to do valuable work and work that is somehow theirs! But they are thwarted.

Is not this the "waste of our human resources"?

The case is that by the "sole-prerogative" clause in union contracts the employer has the sole right to determine what is to be produced, how it is to be produced, what plants are to be built and where, what kinds of machinery are to be installed, when workers are to be hired and laid off, and how production operations are to be rationalized. (Frank Marquart.) There is none of this that is inevitable in running a machine economy; but if these are the circumstances, it is not surprising that the factory operatives' actual code has absolutely nothing to do with useful service or increasing production, but is notoriously devoted to "interpersonal relations"; (1) don't turn out too much work; (2) don't turn out too little work; (3) don't squeal on a fellow worker; (4) don't act like a big-shot. This is how to belong.

4.

Let us go on to the Occupational Outlook of those who are verbally bright. Among this group, simply because they cannot help asking more general questions — e.g., about utility — the problem of finding man's work is harder, and their disillusion is more poignant.

• • •

The more intelligent worker's "indifference" is likely to appear more nakedly as profound resignation, and his cynicism may sharpen to outright racketeering.

"Teaching," says the <u>Handbook</u>, "is the largest of the professions." So suppose our new verbally bright young man chooses for teacher, in the high school system or, by exception, in the elementary schools if he understands that the elementary grades are the vitally important ones and require the most ability to teach well (and of course they have less prestige). Teaching is necessary and useful work; it is real and creative, for it directly confronts an important subject matter, the children themselves; it is obviously self-justifying; and it is ennobled by the arts and sciences. Those who practice teaching do not for the most part succumb to cynicism or indifference — the children are too immediate and real for the teachers to become callous — but, most of the school systems being what they are, can teachers fail to come to suffer first despair and then deep resignation? Resignation occurs psychologically as follows; frustrated in essential action, they nevertheless cannot quit in anger, because the task is necessary; so the anger turns inward and is felt as resignation. (Naturally, the resigned teacher may then put on a happy face and keep very busy.)

For the job is carried on under impossible conditions of overcrowding and saving public money. Not that there is not enough social wealth, but first things are not put first. Also, the school system has spurious aims. It soon becomes clear that the underlying aims are to relieve the home and keep the kids quiet; or, suddenly, the aim is to produce physicists. Timid supervisors, bigoted clerics, and ignorant school boards forbid real teaching. The emotional release and sexual expression of the children are taboo. A commercially debauched popular culture makes learning disesteemed. The academic curriculum is mangled by the demands of reactionaries, liberals, and demented warriors. Progressive methods are emasculated. Attention to each case is out of the question, and all the children — the bright, the average, and the dull — are systematically retarded one way or another, while the teacher's hands are tied. Naturally the pay is low — for the work is hard, useful, and of public concern, all three of which qualities tend to bring lower pay. It is alleged that the low pay is why there is a shortage of teachers and why the best do not choose the profession. My guess is that the best avoid it because of the certainty of miseducating. Nor are the best wanted by the system, for they are not safe. Bertrand Russell was rejected by New York's City College and would not have been accepted in a New York grade school.

5.

Next, what happens to the verbally bright who have no zeal for a serviceable profession and who have no particular scientific or artistic bent? For the most part they make up the tribes of salesmanship, entertainment, business management, promotion, and advertising. Here of course there is no question of utility or honor to begin with, so an ingenuous boy will not look here for a manly career. Nevertheless, though we can pass by the sufferings of these well-paid callings, much publicized by their own writers, they are important to our theme because of the model they present to the growing boy.

Consider the men and women in TV advertisements, demonstrating the product and singing the jingle. They are clowns and mannequins, in grimace, speech, and action. And again, what I want to call attention to in this advertising is not the economic problem of synthetic demand, and not the cultural problem of Popular Culture, but the human problem that these are human beings working as clowns; that the writers and designers of it are human beings thinking like idiots; and the broadcasters and underwriters know and abet what goes on —

> Juicily glubbily
> Blubber is dubbily
> delicious and nutritious
> — eat it, Kitty, it's good.

Alternately, they are liars, confidence men, smooth talkers, obsequious, insolent, etc., etc.

The popular-cultural content of the advertisements is somewhat neutralized by Mad magazine, the bible of the twelve-year-olds who can read. But far more influential and hard to counteract is the fact that the workmen and the patrons of this enterprise are human beings. (Highly approved, too.) They are not good models for a boy looking for a manly job that is useful and necessary, requiring human energy and capacity, and that can be done with honor and dignity. They are a good sign that not many such jobs will be available.

The popular estimation is rather different. Consider the following: "As one possible aid, I suggested to the Senate subcommittee that they alert celebrities and leaders in the fields of sports, movies, theater and television to the help they can offer by getting close to these [delinquent] kids. By giving them positive 'heroes' they know and can talk to, instead of the misguided image of trouble-making buddies, they could aid greatly in guiding these normal aspirations for fame and status into wholesome progressive channels." (Jackie Robinson, who was formerly on the Connecticut Parole Board.) Or again: when a mass cross-section of Oklahoma high school juniors and seniors was asked which living person they would like to be, the boys named Pat Boone, Ricky Nelson, and President Eisenhower; the girls chose Debbie Reynolds, Elizabeth Taylor, and Natalie Wood.

The rigged Quiz shows, which created a scandal in 1959, were a remarkably pure distillate of our American cookery. We start with the brute facts that (a) in our abundant expanding economy it is necessary to give money away to increase spending, production, and profits; and (b) that this money must not be used for useful public goods in taxes, but must be plowed back as "business expenses," even though there is a shameful shortage of schools, housing, etc. Yet when the TV people at first tried simply to give the money away for nothing (for having heard of George Washington), there was a great Calvinistic outcry that this was demoralizing (we may gamble on the horses only to improve the breed). So they hit on the notion of a real contest with prizes. But then, of course, they could not resist making the show itself profitable, and competitive in the (also rigged) ratings with other shows, so the experts in the entertainment-commodity manufactured phony contests. And to cap the climax of fraudulence, the hero of the phony contests proceeded to persuade himself, so he says, that his behavior was educational!

The behavior of the networks was correspondingly typical. These business organizations claim the loyalty of their employees, but at the first breath of trouble they were ruthless and disloyal to their employees. (Even McCarthy was loyal to his gang.) They want to maximize profits and yet be absolutely safe from any risk. Consider their claim that they knew nothing about the fraud. But if they watched the shows that they were broadcasting, they could not possibly, as professionals, not have known the facts, for there were obvious type-casting, acting, plot, etc. If they are not professionals, they are incompetent. But if they don't watch what they broadcast, then they are utterly irresponsible and on what grounds do they have the franchises to the channels? We may offer them the choice: that they are liars or incompetent or irresponsible.

The later direction of the investigation seems to me more important, the inquiry into the bribed disk-jockeying; for this deals directly with our crucial economic problem of synthesized demand, made taste, debauching the public and preventing the emergence and formation of natural taste. In such circumstances there cannot possibly be an American culture; we are doomed to nausea and barbarism. And then these baboons have the effrontery to declare that they give the people what the people demand and that they are not responsible for the level of the movies, the music, the plays, the books!

Finally, in leafing through the Occupational Outlook Handbook, we notice that the armed forces employ a large number. Here our young man can become involved in a world-wide demented enterprise, with personnel and activities corresponding.

6.

Thus, on the simple criteria of unquestioned utility, employing human capacities, and honor, there are not enough worthy jobs in our economy for average boys and adolescents to grow up toward. There are of course thousands of jobs that are worthy and self-justifying, and thousands that can be made so by stubborn integrity, especially if one can work as an independent. Extraordinary intelligence or special talent, also, can often carve out a place for itself — conversely, their usual corruption and waste are all the more sickening. But by and large our economic society is not geared for the cultivation of its young or the attainment of important goals that they can work toward.

This is evident from the usual kind of vocational guidance, which consists of measuring the boy and finding some place in the economy where he can be fitted; chopping him down to make him fit; or neglecting him if they can't find his slot. Personnel directors do not much try to scrutinize the economy in order to find some activity that is a real opportunity for the boy, and then to create an opportunity if they can't find one. To do this would be a horrendous task; I am not sure it could be done if we wanted to do it. But the question is whether anything less makes sense if we mean to speak seriously about the troubles of the young men.

Surely by now, however, many readers are objecting that this entire argument is pointless because people in fact don't think of their jobs in this way at all. Nobody asks if a job is useful or honorable (within the limits of business ethics). A man gets a job that pays well, or well enough, that has prestige, and good conditions, or at least tolerable conditions. I agree with these objections as to the fact. (I hope we are wrong.) But the question is what it means to grow up into such a fact as: "During my productive years I will spend eight hours a day doing what is no good."

7.

Yet, economically and vocationally, a very large population of the young people are in a plight more drastic than anything so far mentioned. In our society as it is, there are not enough worthy jobs. But if our society, being as it is, were run more efficiently and soberly, for a majority there would soon not be any jobs at all. There is at present nearly full employment and there may be for some years, yet a vast number of young people are rationally unemployable, useless. This paradox is essential to explain their present temper.

Our society, which is not geared to the cultivation of its young, is geared to a profitable expanding production, a so-called high standard of living of mediocre value, and the maintenance of nearly full employment. Politically, the chief of these is full employment. In a crisis, when profitable production is temporarily curtailed, government spending increases and jobs are manufactured. In "normalcy" — a condition of slow boom — the easy credit, installment buying, and artificially induced demand for useless goods create jobs for all and good profits for some.

Now, back in the Thirties, when the New Deal attempted by hook or crook to put people back to work and give them money to revive the shattered economy, there was an outcry of moral indignation from the conservatives that many of the jobs were "boondoggling," useless made-work. It was insisted, and rightly, that such work was demoralizing to the workers themselves. It is a question of a word, but a candid critic might certainly say that many of the jobs in our present "normal" production are useless made-work. The tail fins and built-in obsolescence might be called boondoggling. The $64,000 Question and the busy hum of Madison Avenue might certainly be called boondoggling. Certain tax-dodge Foundations are boondoggling. What of business lunches and expense accounts? fringe benefits? the comic categories of occupation in the building trades? the extra stagehands and musicians of the theater crafts? These jolly devices to put money back to work no doubt have a demoralizing effect on somebody or other (certainly on me, they make me green with envy), but where is the moral indignation from Top Management?

Suppose we would cut out the boondoggling and gear our society to a more sensible abundance, with efficient production of quality goods, distribution in a natural market, counterinflation and sober credit. At once the work week would be cut to, say, twenty hours instead of forty. (Important People have already mentioned the figure thirty.) Or alternately, half the labor force would be unemployed. Suppose too — and how can we not suppose it? — that the automatic machines are used generally, rather than just to get rid of badly organized unskilled labor. The unemployment will be still more drastic.

(To give the most striking example: in steel, the annual increase in productivity is 4 per cent, the plants work at 50 per cent of capacity, and the companies can break even and stop producing at <u>less</u> <u>than</u> <u>30</u> <u>per</u> <u>cent</u> of capacity. These are the conditions that forced the steel strike, as desperate self-protection. Estes Kefauver, quoting Gardiner Means and Fred Gardner.)

Everybody knows this, nobody wants to talk about it much, for we don't know how to cope with it. The effect is that we are living a kind of lie. Long ago, labor leaders used to fight for the shorter work week, but now they don't, because they're pretty sure they don't want it. Indeed, when hours are reduced, the tendency is to get a second, part-time, job and raise the standard of living, because the job is meaningless and one must have something; but the standard of living is pretty meaningless, too. Nor is this strange atmosphere a new thing. For at least a generation the maximum sensible use of our productivity could have thrown a vast population out of work, or relieved everybody of a lot of useless work, depending on how you take it. (Consider with how little cutback of useful civilian production the economy produced the war goods and maintained an Army, economically unemployed.) The plain truth is that at present very many of us are useless, not needed, rationally unemployable. It is in this paradoxical atmosphere that young persons grow up. It looks busy and expansive, but it is rationally at a stalemate.

8.

These considerations apply to all ages and classes; but it is of course among poor youth (and the aged) that they show up first and worst. They are the most unemployable. For a long time our society has not been geared to the cultivation of the young. In our country 42 per cent have graduated from high school (predicted census, 1960); less than 8 per cent have graduated from college. The high school trend for at least the near future is not much different: there will be a high proportion of drop-outs before the twelfth grade; but markedly more of the rest will go on to college; that is, the stratification will harden. Now the schooling in neither the high schools nor the colleges is much good — if it were better more kids would stick to it; yet at present; if we made a list we should find that a large proportion of the dwindling number of unquestionably useful or self-justifying jobs, in the humane professions and the arts and sciences, require education; and in the future, there is no doubt that the more educated will have the jobs, in running an efficient, highly technical economy and an administrative society placing a premium on verbal skills.

(Between 1947 and 1957, professional and technical workers increased 61 per cent, clerical workers 23 per cent, but factory operatives only 4½ per cent and laborers 4 per cent. — Census.)

For the uneducated there will be no jobs at all. This is humanly most unfortunate, for presumably those who have learned something in schools, and have the knack of surviving the boredom of those schools, could also make something of idleness; whereas the uneducated are useless at leisure too. It takes application, a fine sense of value, and a powerful community-spirit for a people to have serious leisure, and this has not been the genius of the Americans.

From this point of view we can sympathetically understand the pathos of our American school policy, which otherwise seems to be inexplicable; at great expense compelling kids to go to school who do not want to and who will not profit by it. There are of course unpedagogic motives, like relieving the home, controlling delinquency, and keeping kids from competing for jobs. But there is also this desperately earnest pedagogic motive, of preparing the kids to take some part in a democratic society that does not need them. Otherwise, what will become of them, if they don't know anything?

Compulsory public education spread universally during the nineteenth century to provide the reading, writing, and arithmetic necessary to build a modern industrial economy. With the overmaturity of the economy, the teachers are struggling to preserve the elementary system when the economy no longer requires it and is stingy about paying for it. The demand is for scientists and technicians, the 15 per cent of the "academically talented." "For a vast majority [in the high schools]," says Dr. Conant in The Child, the Parent, and the State, "the vocational courses are the vital core of the program. They represent something related directly to the ambitions of the boys and girls." But somehow, far more than half of these quit. How is that?

9.

Let us sum up again. The majority of young people are faced with the following alternative: Either society is a benevolently frivolous racket in which they'll manage to boondoggle, though less profitably than the more privileged; or society is serious (and they hope still benevolent enough to support them), but they are useless and hopelessly out. Such thoughts do not encourage productive life. Naturally young people are more sanguine and look for man's work, but few find it. Some settle for a "good job"; most settle for a lousy job; a few, but an increasing number, don't settle.

I often ask, "What do you want to work at? If you have the chance. When you get out of school, college, the service, etc."

Some answer right off and tell their definite plans and projects, highly approved by Papa. I'm pleased for them, but it's a bit boring, because they are such squares.

Quite a few will, with prompting, come out with astounding stereotyped, conceited fantasies, such as becoming a movie actor when they are "discovered" — "like Marlon Brando, but in my own way."

Very rarely somebody will, maybe defiantly and defensively, maybe diffidently but proudly, make you know that he knows very well what he is going to do; it is something great; and he is indeed already doing it, which is the real test.

The usual answer, perhaps the normal answer, is "I don't know," meaning, "I'm looking; I haven't found the right thing; it's discouraging but not hopeless."

But the terrible answer is, "Nothing." The young man doesn't want to do anything.

— I remember talking to half a dozen young fellows at Van Wagner's Beach outside of Hamilton, Ontario; and all of them had this one thing to say: "Nothing." They didn't believe that what to work at was the kind of thing one wanted. They rather expected that two or three of them would work for the electric company in town, but they couldn't care less. I turned away from the conversation abruptly because of the uncontrollable burning tears in my eyes and constriction in my chest. Not feeling sorry for them, but tears of frank dismay for the waste of our humanity (they were nice kids). And it is out of that incident that many years later I am writing this book.

Adolescent Drug Dependence: An Expression of an Urge

Roy Goulding

Summary

Now it is of course presumptuous of me, or of anyone else, to suggest that I could sum up adequately for each and all of you at the conclusion of this Symposium, because you have been presented with a kaleidoscope of expert opinion. Even if I take the highlights as I see them, they will not necessarily be the highlights that have attracted you. But, at least apart from the President, I have the last word and I can escape perhaps just with my skin.

I understand – and this is quite clear – that the Symposium that is now being concluded is in fact the first of its kind organized by the Society and most of you, being members, will realize this. But, for those of you who are visitors and do not, this Society is essentially a scientific body with an international membership, having for its objects – and I stress this – "the systematic study of addiction to alcohol and other drugs, and the investigation of all forms of alcoholism." The Society does not seek to exercise any control over the opinions or practice of its members. Of course, with the reception afterwards in mind, that is rather a relieving thought!

*Adapted from the Concluding Remarks by Roy Goulding, Proceedings of the Society for the Study of Addiction, London, 1 and 2 September 1966: The Pharmacological and Epidemiological Aspects of Adolescent Drug Dependence, C. W. M. Wilson, Editor, Pergamon Press Publishers. Copyright © 1968. The Society for the Study of Drug Addiction.

At any rate, I think the objects of the Society have been well displayed in these last 2 days, because what has been attempted has been, so far as possible within the limited knowledge available, an objective study of the facts. If, in time, the conclusions drawn from the accumulated facts are used by authorities, governmental and otherwise, to modify and further official policy, so well and good.

Looking over some of the papers and thinking about them, certain items stick very much in one's mind. I think that all of us would be very interested in the idea of dissociation between powerful analgesic action, euphoria and drug dependence. This is a quest which many have pursued for years and which now looks like being successful. Of course, that will not necessarily relieve us of the problem of addiction generally. It is only a small proportion of our addicts today that are therapeutic in their initiation and those who are not therapeutic addicts do not take drugs in the first place because they are analgesic. Not that this should for one moment deter the chemists in their progress along these lines.

I think the point about the ultimate prognosis and "maturation", as it were, of the addict, lends force to the need for a long-term research and I am not at all sure that already in this country we are possibly losing an opportunity. We now have, regrettably, a number of young addicts who have been on the drugs 2, 3, 4 or 5 years and it is high time that we decided to keep a surveillance over them, not in a castigory sense, but purely in the research sense, to see how they develop and what happens to them in the end.

We were shown, with work on animals, the experimental approach. This is something which I find very attractive. But I am also worried about it. In the conventional approach to toxicology, to the toxicology of drugs and other materials, we have perforce to resort to animals because, for ethical and other reasons, we cannot plunge straight away into man or, for that matter, into woman either. Now, our great trouble in toxicological research of this kind is that of extrapolating from these animals, no matter how many thousands of them there may be, to man. Many of us have said time and time again that if only we could start our research on man it would be so much more useful and so much more valuable. For this reason I therefore cannot work up this terrible keenness to embark on a lot of animal work, unless it is specifically to explain some problem that we meet clinically that we cannot resolve otherwise. If I may be forgiven I must disagree that all essential facts about drugs have come from animal work; I have a nasty memory about ergot.

The studies with amphetamine addicts is a field in which a tremendous amount of invaluable work has been done already and I like the idea that what is wanted for so many drug addicts is proper out-patient supervision. If, indeed, certain of the recommendations of the Brain Committee, Mark II, are to be implemented, then it will not be sufficient just to say that doctors generally cannot give drugs to addicts, but that the addicts must go to the hospital. This can easily turn the hospital into a sort of slot-machine taking the place of the general practitioner. Hospital centres must be properly equipped and manned so that they are able to keep proper control over these people, provide for investigation and treatment and collaborate with others on a multi-disciplinary basis for furthering the whole study of addiction.

When it comes to the anorexogenic drugs as I think they are called, then I think a good time can be had by all and obviously a jolly good time is going to be had by all when the laxatives are included as well! I cannot help thinking that, although it was stressed that obesity is essentially a function of diet, one of the strongest elements in weight reduction is, in fact, bowel action. One has only got to observe patients suffering from ulcerative colitis for a few days to see how they waste away. So, if they put a laxative with the phenmetrazine, they will get the biggest weight loss they have seen in centuries!

Here at this gathering we have had a very large team from Sweden and I for one am very glad to see them here. We are always pleased to receive them. I never cease to wonder why they get anything like the epidemic of phenmetrazine dependence such as we have heard them describe. It puzzles me all the more in a society that is so prosperous, that has so much leisure and so many noble urges. I suppose it is just a matter of the way in which those urges become socially channelled.

We are all very worried about LSD, mescaline and the like. I think it raises a very awkward problem officially, if I may look at it from that angle for a moment. It would be difficult to get universal opinion that drugs of this sort are of no use at all in the practice of psychiatry. Some of us would like to say so, but there are those who would differ. Consequently, a complete and utter ban on such-like drugs would not meet with professional approval. One is left therefore with the necessity of saying that only bona-fide uses of LSD should be condoned, but that all others should be excluded. Now, who are the "right" people to be entrusted with it? Are you here going to be approved, and you there be rejected? This is an awfully difficult thing to do. In this country we are, I understand, confronted with this very decision at the present time and I do not quite know how it should be resolved.

I am always absorbed by the accounts, most of which I read in the more sensational papers that I find on the common room tables — naturally I do not buy them — the accounts of juvenile, and, particularly, undergraduate drug taking and I think it is very important that people who are able to study this first-hand should be able to give us as accurate an estimate as possible of the overall incidence of this. Otherwise, we do tend to form the impression that the whole of adolescent society is decaying in some sort of morass of opiates, marijuana and similar drugs. I think — and in so doing I hope that I won't be accused of complacency — I think the proportion is very small. I'm not dismissing it for a moment, but I think it is important that we should not get this out of proportion.

The comments on drug taking, medicine imbibing, and suchlike being a feature of life nowadays are very close to my own heart. I think very definitely that medicine taking has gone too far. We live, I gather, in a more and more materialistic age and the general opinion, aided by the advertisements, mass media and so on, is for every slight deviation from health, physical or mental, to be treatable by some scientific, chemical or biochemical corrective. Salvation lies in drugs, so the public is led to believe. Consequently, it is looked upon as the sort of "done thing" to take one's chemical corrective, as Noel Coward put it "to knock us in, or knock us out of shape". This has become such a part of our daily lives that I think too many of the youngsters grow up being given a couple of aspirins whenever they feel a bit headachy, a couple of something else when they feel a bit disturbed inside and a couple of this, that or the other when they feel a bit disagreeable, to such an extent that they are unaware of crossing any great gulf when they embark on drug taking for a "lift". It is, however, difficult to understand why, at that stage and against that background, some experiment and do not go on, whereas others become completely "hooked".

Now, I personally have become almost sickened on hearing the accounts of the case histories of young drug addicts to hear that they come from broken homes — father wasn't always there; mother was possessive; brother did this, that or the other; there was a big emotional element. There was a devil of an emotional element in my family, but, as far as I know — and you have got to take my word for this — I'm not "hooked" yet! I'm glad I was subjected to emotional stresses. I didn't want to come from a phlegmatic home. I think we have got to be very careful, when we do a thorough check-up on the backgrounds of adolescent drug takers, that we pay equal attention to the family backgrounds of those who took the stuff and didn't keep on it, to see where the differences comes in.

My own feeling is that there are not just addicts; there are different forms of addicts. And by talking about adolescent addicts we are generalizing at the risk of obscuring a number of details and differences which are very important for the further prevention and treatment.

Those of you from overseas will have seen the horrid statistical confirmation of the state of affairs that has caused us here quite a lot of worry recently. I think we have all got worked up, and quite rightly so, about all these extra heroin and heroin-cocaine addicts. We are disturbed, though not quite worked up, about those on cannabis. The amphetamine victims some people seem to disregard and as for the barbiturates, oh, we almost couldn't care less about them. At the same time, the alcoholics as it was quite rightly stressed, still give rise to more trouble, suffering, upsets and so on, than probably all the rest put together.

One thing I forgot when I was talking about the family background and case histories was the very important and, to my mind, very pertinent point about patient testimony. It is very difficult to get independent evidence and confirmation. But drug addicts, being the sort of people they are, like to tell a good story and they like, moreover, to tell a story if this helps to excuse their own guilt. So, it was because "mother made too much fuss over them", or "father gave them too much of the strap", that they got in with the wrong lot, when probably circumstances viewed from outside would suggest a rather differently slanted state of affairs.

About the pamphlet, a sort of do-it-yourself book on drug taking and how to prevent it; I don't know who is going to write it if you start sponsoring one and don't accept the library that has already been suggested to you. But if one man writes it — my goodness, he's in for trouble isn't he! He'll please nobody. So you had better set up a committee; then it will never be written at all!

We were given a very revealing account of the findings at Oxford following upon a most careful survey and I, for one, am relieved (though not an Oxford man myself) to find that not everything is wrong in the State of Oxford. Again the point was made of the relative incidence of this sort of deviation among the large population at our oldest university and, of course, Oxford always attracts the news because it is antiquarian, it is the oldest university in this country. I think it is very interesting that there, in what was supposed to have been a real stronghold of addiction, the number actually confirmed out of the total was in the end so small.

It is curious to see the phases through which undergraduates go and what is, as he said, "the big thing to do". At the moment we are worried that drug taking has become the "done thing". I remember a scout in Cambridge who was asked what difference he recognized between the undergraduate in the 1950's and the one of 50 years ago. He replied: "When I started, Sir, they had a clean shirt every day and a bath once a term. Now, Sir, they have a bath every day and a clean shirt once a term." I gather that now they have neither. But that may be Oxford! I would agree every time and I am sure there will be few people opposing the urge to have stronger sanctions on the pushers.

When we come to Dr. H's contribution, as he himself has stated, he is worried about assuming the nickname of "Lurid Milton" on account of the most impressive pictures which he showed us and the even more impressive accounts that he gave us of what has happened in New York. As the medical examiner he sees the end result. I know that some have been worried whether we might be missing some heroin deaths in this country and attributing them to other causes, to natural diseases, for example, when in fact the cause is drug addiction. But if the number of deaths of this sort recognized by the pathologist is any guide, or index, to the total number of addicts, then perhaps this suggests that our number of addicts is, after all, more or less what has been suggested by the Home Office returns. Oh, and while I mention the Home Office, can I please emphasize once again that we have no register. Addicts are not "registered" in this country. Could I please have this put on record?

The acute, fatal reactions that have been described are fascinating to me as they are to many others, particularly pathologists. With a reaction like the flooding of the lungs, one thinks immediately of some acute allergic phenomenon, just as one sees in experimental anaphylaxis. Now this, I suppose, is just a frivolous suggestion, but one thing that occurred to me is this. For many years we have had our share of addicts here; not all that number, perhaps, but a quota of regular heroin takers, nevertheless, who administer the pure stuff, or as pure as our manufacturers can contrive it. Our lot, though, always tend to put it in with a real syringe and plunger. Yours, on the other hand, put it in with a rubber-ended contrivance. An awful lot of funny things can come out of rubber and I wonder whether repeated doses can cause a sensitization reaction from this source.

What I really find myself preoccupied with is the philosophical aspect. I am virtually a non-smoker, but not simply because I believe that in this way I will develop lung cancer. It is just that smoking makes me cough and choke — good cigars apart, of course. But we warn everybody pretty impressively, as far as I can see, about getting cancer from cigarette smoking. If, then, they choose to go to the devil, I don't see why we should try to stop them. And if somebody really wants to be a drug addict, are we justified in stopping him so long as he is no trouble to anybody else? If it becomes a contagion, then something must be done, but I think we must be very, very careful in our motives here, so as not to confuse protection of the community with putting right the individual who doesn't necessarily want to be put right anyway.

The inevitable argument arose over the compulsory versus voluntary treatment in management. As the discussion showed, it is not a clear-cut distinction as to whether you put them inside with iron bars around them, or to let them go entirely free. It is how long you keep them in with some restraint and how long you support them afterwards, with voluntary co-operation and so on. This must be separately tailored to the different cases. For some individuals we must be prepared to go to one extreme, for others to the opposite extreme and for the greater mass of addicts to take a middle course. Obviously treatment is not merely a matter of getting addicts off their drugs, it is not merely a matter of getting them walking about again, but it is the task of getting them back as useful members of society and back in a profitable occupation.

This talk of profitable occupation led to mention of the "age of leisure" — one month on work and one month off, so to speak. Well, I was a bit worried with L's suggested athletic programme. It really shook me to the core. I have recently returned from a conference in Europe about drug taking by athletes. It seems to me that he is just creating the opportunity for it. I don't want any imposed athletic programmes. You must forgive me, but I really was worried by the United States of America, of all countries, talking about stipulating government guide lines for our leisure. I'm off to Peking — there may be more fun there!

Well, I cannot claim to have studied the leisured population here, though it is my belief that when people are "off work" they are often occupied much more energetically than when they are in the factory. They are rushing around with all this "do-it-yourself" business nowadays — mend your own car, paint your own house, dig your own garden, build your own boat. You then see the fellow going back to work on Monday morning for a rest! We are not perturbed about the leisure. Seriously, though, what I am referring to is the adult. It is not quite the same with the adolescent. The "do-it-yourself" attracts a family man, or it is enforced upon him by the family. But the youngsters may not be quite so enthusiastic in this way. Quite a lot of them, nevertheless, are. It does, however, seem important to me that we must consider interesting young people in something that is a little dangerous, daring, risky and so on — something of this sort, on which they can embark in their spare time rather than let them look for outlets in the form of drug taking.

The possible connection between crime and addiction was fully considered. American experience in this field is well known. But in Britain I think the position is rather different. Most of the addicts can get the drugs they want, the "hard drugs" anyway, without resort to illicit dealing and they are not prompted to engage in crime, petty or otherwise, in order to get the wherewithall for their supplies.

What is important to all of us is further research. Lots of proposals have been advanced both from the platform and from the body of the hall. So it is really unnecessary for me to enlarge upon this, though there are one or two points to which I might refer once again. There is the question of the toxicological detection and assay of drugs in body fluids. In this I have a personal interest. We are hopeful of getting a laboratory going within the next few months, though not expressly for this purpose. Our object is to work out simple methods and to provide a service for finding and estimating drugs in cases of acute poisoning. Obviously addiction may come into this as well.

As for research, there is so much to be done and there is so much (to use that horrible word when talking about fellow human beings) "clinical material" to hand. We need not look for contrived experiments; we can do an awful lot of field work to good effect with the material already available.

To conclude with just a comment or two of my own: I said earlier that I thought that to talk of drug addicts as a generalization was an imprecise expression. There are all sorts of separate individuals who become dependent on a variety of drugs. These drugs have different actions.

To take the barbiturates, for example; they vaguely sedate people, make them sleepy and even render them more easy to live with. Apart from acute poisoning (and I am not dealing with that now), they seem to do very little harm. Indeed, socially and individually it is conceivable that overall they do more good than otherwise. Then there are the tranquillizers (and this, I admit, is a very loose term). All sorts of drugs come into this category but, from the point of view of dependence, they are probably of a similar status to the barbiturates.

When, however, we look upon the antidepressives, the amphetamines and such-like substances, we are confronted with a very different state of affairs. These, of course — as everyone here has agreed — have a very distinct pharmacological action. They are stimulants or excitants. On this account they exercise a definite attraction in themselves. For this reason alone they probably merit a much more rigid statutory control. They have a greater potential in themselves for misuse and harm.

Even worse for their intrinsic effects are the narcotics — opium, morphine, heroin and cocaine as well. These have an enormous capacity for addiction. Hence, the specially rigid restrictions applied to them. The victims to these drugs are probably in the most desperate situation of all.

Alcohol, then, is in a class by itself. No one here, I trust, will deny that it has charms.Chronic over-indulgence can lead to severe degradation. Statutorily, however, we do little to control its supply, except withholding it from children. Yet drug of dependence it surely is, but one that we regard with almost over-complacent tolerance. The alcoholic, moreover, may be of quite a different personality from the morphomaniac.

Finally we have LSD and mescaline, both undoubtedly terribly dangerous drugs. They justify the most determined supervision. But whether they really give rise to dependence is open to question.

This Symposium is, after all, devoted to adolescents. We are dealing with a class of the community that is growing up, and raring to get going. It wants to venture into new territories, undergo novel experiences and, if there is a risk attached to them, so much the better. The youngster who doesn't want to do anything at all of this kind is surely a very unpromising specimen after all. These youngsters find various ways to express themselves. Some of them construct the most elaborate scale models; they are the obsessive types. They get worked up if things don't go right. They are meticulously correct. Others, as we have heard today from the platform, tear about on motor-bikes. They are the "ton-up" boys with the leather jackets. A few (and this has been our interest here today and yesterday) take to drugs for the sheer excitement of it. We get very anxious about them, we think that they are doing themselves harm and that they are doing harm to others as well. So we think about ways of preventing these antics, of stopping them and so on. Yet one doesn't have to have motor-bikes or dangerous drugs to get into trouble. For instance, one can indulge in group fetishes, mass bullying, or vandalism at various seaside towns. These are all the acts of people making a burst-out, a troublesome burst-out, and a burst-out that is unacceptable to the community.

I am reminded here of the history of medicine and of the time, a century or so ago, when the practitioner concentrated his attention on curing the fever, the pain and so on. He was treating, quite rightly in a way, the symptoms. He wasn't curing the disease. Drug taking, it seems to me, is only an expression of an urge, an excessive urge in some people, a lesser urge in others, and an absent urge in a further group. The biggest research effort that can be applied will surely be to study all of these people socially and psychologically, not only the ones who burst out, but also the ones who do not burst out. We should ascertain why they have this irresistible urge and why for some it takes the form of drugs, for others motor-bikes, while others lean towards sex. But anyway, there is a deviation of one sort or another which is an expression of an urge that is not otherwise reasonably satisfied.

Mr. President, I think your Society has done an honour to science and to the study of drug dependence generally in so successfully organizing this Symposium. I, for one, have enjoyed it enormously and in so saying I feel without hesitation that I am voicing the feelings of all the rest of you. Thank you for having brought so many people together, for letting us hear so many opinions and for sending us away with a determination to do something about it.

The Nervous System and its Organization: An Essay in Five Parts

1. The gross structure of the brain

In an earlier discussion of organ transplantations I said, "Transplant my head and I go with my head." By this I meant that the individual person represents a collection of accumulated experiences and is identified by means of understandable interactions with other individuals as well as by his behavior in a variety of situations; hence "person" is synonymous with mind. A person's body can change with age and parts of it may even be discarded or replaced; nevertheless, the one aspect of a person that ties together the past and the present, that makes plans and promises for the future is that person's mind. For all practical purposes, the mind resides in the brain.

The brain, of course, is not self-sufficient. It does not exist by itself as an entity separate from other organs and organ systems. No, indeed. About one-eighth of man's circulation is assigned the task of serving his brain. Furthermore, the sense organs — touch, taste, sight, smell, and hearing — exist to feed information about one's surroundings to the brain; if the information thus received calls for action, this action (except in the case of the simplest reflexes) is decided upon in the brain. The brain runs the body.

The brain is a wonderfully complex computer. There are some people who would question this statement by referring to the many abstractions of which human thought is capable. At one time, too, these people would have cited the inability of computers to perform tasks for which they were not programmed. That argument was shattered — almost as it was being advanced — by the development of computers that learn through past errors, by chess-playing computers and their ilk.

By "wonderfully complex computer" I mean one whose complexity has not been approached by man-made models. The optic nerve, which contains the individual fibers connecting the retina of the eye with the visual centers of the brain, has about 1 million nerve fibers. Were these fibers represented by the fine insulated wire used by hobbyists or telephone repairmen, the optic nerve would

261

be a cable roughly three feet in diameter. We cannot yet begin to compare man-made machines and brains. On the other hand, to explain the mind there is no need to look beyond the organization of the brain and its associated parts (including the chemical molecules that are involved in controlling the transmission of signals from one nerve cell to another).

This essay touches on the gross physical aspects of the brain. It will oscillate between the simple and the complex: the simple brain of lower animals and the complex brain of man, the simple brain of an embryo and the complex brain of an adult, and the simple brain of a prehistoric horse and the complex brain of the modern pony. It will introduce in at least a crude way the relationship between the physical organization of the brain and what we regard as its function. This treatment of the brain is necessary in order to develop a basis for understanding sensations, their origins, their interpretation and their role in behavior.

The central nervous system — the brain and the spinal cord — arises in the very young embryo (see the accompanying diagram) as an infolding of the outer layer of cells along what will become the back of the developing individual. The folding forms a groove; the lips of the groove meet and fuse thus forming a hollow tube of cells lying just below the "skin" of the embryo:

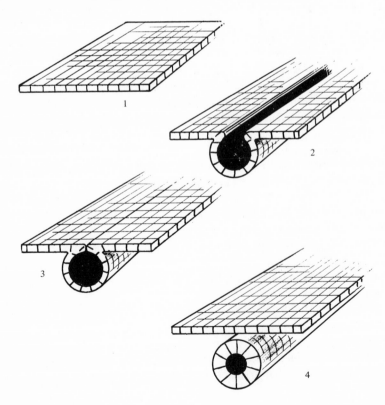

This hollow tube (neural tube) becomes the central nervous system. In the adult, vesicles in the brain and a canal in the spinal cord are the remains of the hollow center of the early neural tube.

A striking feature of active, free-living animals is the localization of numerous sense organs at the anterior or head end of the individual. This is the end that first encounters unexplored territory as the animal moves about. Here, for instance, are the eyes, nose, and ears of mammals; here, too, is the mouth and its assortment of taste buds. Wherever nerves are numerous, wherever subsidiary nerves are needed to interconnect a host of incoming sensory nerves with outgoing motor nerves, the nervous system must enlarge because nerves require space. Hence, the formation of a brain at the anterior end of the neural tube. It is an enlarged region within which an especially large number of nerve impulses are received, analyzed, and shunted to still other nerves in order to evoke appropriate reactions from the body's muscular system: flee or fight, zig or zag, swallow or spit out, court and mate or ignore. Even lowly worms have a constant barrage of messages arriving in their brains that must be interpreted and acted upon.

A survey of brains of vertebrate animals (animals that possess backbones) from fish to man reveals a sequence that grows increasingly complicated, and yet the correspondence of individual parts of the different brains can be identified quite easily. Thus, in the shark-like fishes the olfactory lobes and another part of the brain known as the cerebellum are exceptionally large. Aside from sight and smell (senses handled by the optic and olfactory lobes of the brain), the main functions of the shark's brain are housed in the cerebellum and the part of the brain just anterior to the spinal cord, the medulla oblongata. The cerebral hemispheres, the large convoluted portion of the mammalian brain, exist in sharks only as slight thickenings of the upper surface of the olfactory lobes.

In the higher, bony fishes the cerebellum and the olfactory lobes are reduced; the optic lobes, however, are large. The bony fishes, like birds, rely a great deal on visual signals and the optic lobes of both the bony fishes and birds form substantial portions of the brains of these animals.

In passing from the amphibia, through reptiles and birds, to mammals there is a progressively greater enlargement of the cerebral hemispheres and, relatively speaking, a corresponding dwindling of the cerebellum. The predominant portion of the brain in shark-like fishes has been displaced in higher animals from its position of dominance by large smooth hemispheres whose precursors are scarcely discernible in the shark's brain. The dominance of the cerebral hemispheres in mammals is reflected in an enlargement and convolution of their surface, which culminates in the human brain where the two hemispheres entirely hide the remainder of the brain if viewed from the top. The optic lobes that are distinct bodies in the brains of lower vertebrates are incorporated into the cerebral hemispheres of mammals and man.

The embryological development of the brain in mammals follows a course surprisingly similar to the sequence observed in the survey of various forms of contemporary life from fish to man. The tendency of developing embryos at successive stages to simulate an evolutionary sequence or a sequence reconstructed by a study of present-day forms is a common one. Biologists say that *ontogeny recapitulates phylogeny*. This claim is not to be taken literally. A human embryo does not go through a fish, an amphibian, a reptilian, and then a mammalian stage in approaching final human form. Taken with certain reservations, however, the claim contains much that is true. In meeting each challenge during the course of evolution, changes are wrought by the modification of existing structures rather than by the invention of entirely new structures. These successive developmental modifications must in turn meet constraints imposed by an acceptable period of embryonic and fetal development, and so, as if recapitulating its origins, the developing embryo dashes through an abbreviated and highly modified outline of its evolutionary past.

The evolution of the brain within the horse family during the past 55 million years tells us still more about the origins of the brain's present-day complexity. Fossils of prehistoric horses form a rather complete record beginning with a small, five-toed animal about the size of a fox-terrier. The brains of these fossil animals, of course, are not preserved. A plaster cast of the interior of well-preserved fossil skulls can be made, however, and studied. A series of such casts are shown in the illustration on pages 265-266.

The striking feature revealed by the study of brain casts of fossil horses is the dissimilarity between the brains of primitive horses and those or present-day mammals of comparable size. The brain of Eohippus does not resemble that of a miniature horse or of a small mammal; on the contrary, because of its small, smooth cerebral hemispheres, it more nearly resembles a reptilian brain. Roughly 10 million years later, the brain of Mesohippus (an animal the size of a collie) is still not typical of a mammalian brain; the convolutions in the cerebral hemispheres are too few (the surface of these hemispheres is too smooth), and the hemispheres themselves are too small. Merychippus, an animal that lived 20 or 25 million years ago, has a primitive brain according to the standards of modern mammals.

What does this all mean? It means that mammalian brains were not handed out to mammals during very early stages of mammalian evolution to be passed on subsequently to their descendants. On the contrary, within the separate lines of evolutionary descent the brains of modern mammals enlarged and developed their complex structures independently in response to challenges posed by still other evolving mammals. Only after we have gone far in tracing the horse's line of descent do we see the primitive horse's brain approaching the brain structure of modern horse. This can only mean that is was well into any mammalian line of descent that the cerebral hemispheres lost their originally smooth, reptilian or bird-like appearance and became the enlarged, convoluted structures typical of modern mammals. The evolution of the mammalian brain has been the evolution of computers under circumstances where computers have been pitted against one another for continued survival.

Merychippus
20,000,000 - 25,000,000 years

Mesohippus
48,000,000 years

Eohippus
55,000,000 years

Modern
pony

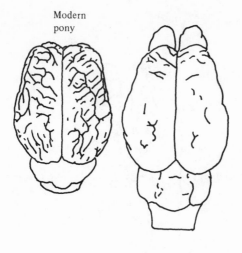

Equus   20,000 years

Equus   1,000,000 years

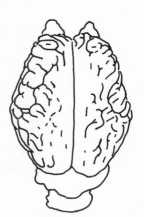

Pliohippus
10,000,000 years

2. The circuitry of the central nervous system

The brain runs the body. Millions of sense organs of various sorts located throughout the body constantly send reports by way of the nerves (in-coming or sensory nerves) to the brain. Because literally millions of reports arrive every instant, the brain must possess mechanisms for setting priorities, for dealing routinely with routine matters, and for meeting extraordinary reports with extraordinary measures. The decisions of the brain are reflected in the messages that again travel to the outlying portion of the body over outgoing (or motor) nerves.

This essay deals in a purely descriptive way with the communication network that is the nervous system. It describes some of the sense organs, the paths of certain nerves leading into and out of the central nervous system, some elementary aspects of nerve impulses, and a very brief account of the nerve paths within the brain itself. The emphasis here is on the central nervous system — the brain and spinal cord — and its role in establishing the relationship of the body to the outside world.

Despite claims to the contrary, or even the misleading appearance of certain individuals, a person's body is never relaxed. No matter in what position he might be — standing, sitting, or lying down — he is under the constant pull of gravity. The weight of his limbs tends to alter the angles of certain joints; his head tends to droop to one side or the other or forward; and his muscles tend to stretch as they relax in surrender to gravitational force. Small sensory receptors that are located in every joint and scattered throughout every skeletal muscle instantaneously detect and report these alterations to the central nervous system. Without the matter ever appearing in his conscious thought, motor signals are sent out to those particular muscles capable of bringing about corrective changes in bodily posture. An after-dinner speaker, for example, continues his talk uninterrupted while hundreds of muscles in his legs and torso tighten and relax in order to keep him balanced on his feet. Meantime, in the audience, an equally complex array of muscles holds each listener's head erect without intruding on the reception of the speech itself.

The nervous pathway described above consists in its simplest form of a sensory receptor, a sensory nerve leading to the central nervous system, a motor nerve leading back to a muscle, and a bundle of muscle fibers. Different receptors detect different aspects of their enrironment. Some encapsulated bodies detect changes in pressure while small skeins of nerve endings report

267

stretchings and other physical distortions; these receptors inform the brain about the physical disposition of the body. Because of these receptors we can bring our finger tips together or touch our toes with our fingers even though our eyes are closed. Some receptors react to changes in temperature; these warn of dangers of heat or cold. The organs of special sense – the eyes, ears, taste buds, and olfactory nerves – have their special functions, which require elaborate and intricate ancillary structures, especially in the case of eyes and ears.

At some point during each physical examination the doctor tests the knee reflex of his patient. He strikes the tendon below the kneecap a sharp blow, one that effectively lengthens the tendon. This lengthening is reported to the spinal cord by distorted receptors. The message is interpreted as meaning that the lower leg has suddenly collapsed backward. Without involving any more of the central nervous system than the spinal cord, an equally abrupt message goes out to those leg muscles that can take corrective action. The result, as everyone knows, is a violent involuntary kick, a reflex action that follows immediately and uncontrollably after the tap on the knee. Following the reflex action, however, is the realization that the knee has been struck, that the body has reacted, and that the reaction was a ridiculous one not under voluntary control. All of these additional sensory ramifications involve more of the nervous system than the simple reflex circuit itself.

What is the nature of the message that is conducted by a nerve? Is there a special message for each stimulus – one for visible light, one for salt, another for pepper, still another for noise, and so forth? Not at all. Each nerve transmits the same type of message. Experimentally, the simplest detectable component of a nerve impulse consists of a sharp spike on a device that is designed to record small electrical currents; the spike reflects a transitory alteration of electrical charge on the outer membrane of a nerve fiber. The magnitude of the alteration for each impulse is constant. The speed with which each impulse travels along the fiber varies a good deal with the size and type of nerve cell but otherwise is constant. The number of impulses that travel along the same fiber in a given interval of time varies with the intensity of stimulus. A nerve, therefore, reports the intensity of a stimulus not by altering the individual impulse nor by altering the speed with which the impulse is transmitted but in increasing the number of impulses that are transmitted in a given time interval.

If all nerves transmit messages in identical form, how can these messages be interpreted properly by the brain? Could sound not be interpreted as light, for example? Yes, indeed. Messages can be misinterpreted. A blow on the eye gives rise to flashes of light; "seeing stars" is no idle phrase. The nerves of the retina, even if stimulated by pressure rather than by light, evoke images of light in the mind. Similarly, loud sounds evoke painful sensations as well as sensations of sound. In addition to errors caused by the false stimulation of receptor organs, the brain at times errs in pinpointing the origin of messages it receives.

This type of error is especially common in respect to pain. The origin of pain is very often referred from one part of the body to another. Many persons are unable to recognize a toothache because the pain appears to be in the throat, upper chest, or even the arm. Heart attacks are often accompanied by pains in the arm. Pain accompanying bladder infections appears to arise well within the chest cavity while infections of the diaphram are at times accompanied by pain in the shoulder.

The correct interpretation of messages received by the brain depends in large part upon their being shunted to a particular part of the brain. In large measure, too, past experience is involved in the interpretation of incoming messages. To an extent greater than we might expect considering the immense complexity of the wiring job itself, the brain is a mosaic in which small sections (and I am thinking now particularly of the convoluted surface of the cerebral hemispheres) are responsible for receiving messages from and returning messages to definite parts of the body. Even the retina of the eye has a crude counterpart in the brain. The retina consists of hundreds of thousands of minute receptors upon which images are thrown by the lens of the eye precisely as the lens of a camera throws an image on the film. As the image moves about on the retina, individual receptors transmit to the brain information on whatever happens to strike them at a given instant. Now, the visual center of the brain is not laid out in the same fine-grain fashion as the retina itself, but there is a rough correlation of groupings of receptors in the eye and corresponding groupings of interpretive centers in the brain. Indeed, the lack of perfect correlation may be necessary if the brain is to distinguish between rotational and straight-line motion, interpret events that are occurring, and anticipate the future position of observed objects.

Neurosurgeons have accumulated a great deal of information about brain damage and physical impairment in patients suffering from strokes and brain tumors. Sensations from one side of the body, for example, are transmitted to the cerebral hemisphere of the opposite side; correspondingly, motor impulses from one hemisphere control the muscles on the opposite side of the body. By noting very carefully both the location of each brain lesion and the accompanying defect in motor and sensory abilities, neurologists have rather accurately mapped the role of different areas of the cerebral surface. The accompanying diagram illustrates in a striking way the absence of correspondence between the size of an organ or region of the body and the area of the brain that is responsible for its control. The torso and legs are controlled by rather small regions of the brain; the muscles concerned with speech, on the other hand, are controlled by a large area. Each muscle that is activated independently of others, even of its immediate neighbors, requires its own controlling site: a set of muscles, each one of which can be activated independently but which are actually called upon to perform complex tasks in cooperation with one another, requires a large controlling center. Hence, the predominance of the motor and sensory areas controlling the hand and fingers in contrast to those controlling the leg, and the predominance of the areas controlling the face, lips, tongue, and throat — the muscles of speech — over all else.

What is the role of experience in the interpretation of messages received by the brain? Here I shall mention three examples that illustrate what I mean when I say that experience does indeed modify the interpretation of messages received in the brain from its sensory receptors. First, without elaboration, I can remind you of optical illusions, which are designed to befuddle us by distortions of usually reliable landmarks. Second, I can remind the amateur photographers among you of the bizarre results you obtain by shooting night photos indoors with an outdoor film. The predominantly orange hues of these photographs are in fact true colors; they appear strange, however, because under indoor light the mind "corrects" the information it receives so that colors of things agree with experience gained during daylight hours. Third, I can mention experiments in which persons are asked to wear glasses equipped with prismatic lenses. Such glasses make the world appear to be upside down. After wearing such glasses for several days, however, the wearer reinverts the world so that it is right side up once more. When the glasses are then removed, the world again appears to be upside down and "rights" itself only after a few additional days. It is also true, of course, that the image of our surroundings that is normally projected upon the retina of each eye is itself inverted.

The relationship of man to his surroundings, in brief, is based entirely upon a series of what are (or appear to be to our mechanical measuring devices) identical electrical impulses that arrive at different parts of the nervous system, only the frequency with which the individual impulses arrive conveys information regarding the intensity or strength of the initiating stimulus. At times the reaction of the body is swift as it is in the case of a burned finger. The arm is jerked back even before the sensation of pain has clearly emerged in conscious thought. These rapid reflexes are controlled directly from the spinal cord. During periods of intense mental concentration all extraneous stimuli from outside are effectively suppressed; a system of assigning priorities to incoming messages is an essential feature of a normally functioning mind. And still, at other times, the mind is left open to receive and absorb a world of diverse stimuli; these are times when sounds, sights, and sensations of touch and smell combine in impressing upon us the beauty of life — as in a stroll along a sea shore. These are times when we speak of "soaking in" our surroundings. Long after the fact, the entire symphony of sensations can be triggered once more by experiencing any one stimulus alone — such is the power of the call of a gull or the sound of the surf.

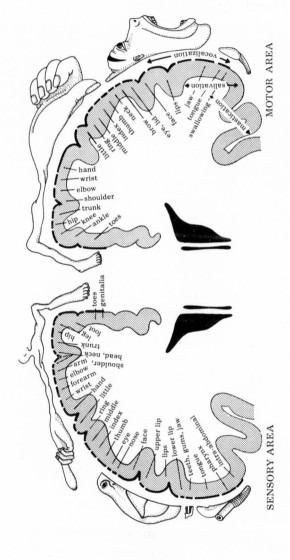

MOTOR AREA

SENSORY AREA

Function maps of the surface of the cerebral cortex of man. [Modified from W. Penfield and T. Rasmussen, The Cerebral Cortex of Man, Macmillan, 1950.]

3. Memory: the storage and recall of information

Compared to the brain of any other vertebrate, that of mammals possesses enormous cerebral hemispheres. Man's cerebrum is the largest, and its surface is the most convoluted of all. The cerebrum integrates sensory information; it "makes sense" out of the multitude of nerve impulses that arrive each instant. Here most of what we have learned is stored. Our speech and our voluntary actions are controlled from nerve centers occupying various sites within the cerebral hemispheres. Conscious thought – during which we mull over alternative possibilities, foresee the outcome of each, and then choose one as a basis for action – is housed in the cerebral hemispheres.

How is all this accomplished? No one knows how experiences are stored in the brain nor how they are recalled as memories; at least, no one knows these things in basic molecular terms. A great deal is known, however, about the physical structure of the brain and the role the cerebral hemispheres play in determining whether experiences will be recorded and, if they are, where these records are stored. The precise content of file drawers may not be understood, but the disposition of filing cabinets has been worked out quite well.

There are two cerebral hemispheres, a left one and a right one. The two are mirror images of one another. They are interconnected at various points by transverse bundles of nerve fibers – about eight bundles in all. The transverse nerve fibers connect corresponding parts of the two hemispheres; the largest band of fibers is known as the great cerebral commissure (or the *corpus callosum*). These commissures are large and obviously important. Nevertheless, early neurosurgeons found to their amazement that they could be severed during operations on the brain with no apparent effect on the patient. What, then, is the role of these transverse connections? In answering that question, we learn a good deal about the disposition of the filing cabinets, that is, of the gross aspects of information storage and recall in the brain.

Man, like most higher organisms, is bilaterally symmetrical; he has two eyes, two ears, two arms, and two legs – left and right in each case. Unpaired organs are found on the midline. The brain is no exception. The two sides of the brain are symmetrical; each has its optic lobe, its olfactory center, its cerebral hemisphere, and so on. This is not to say that the nervous system is a divided structure, for it is not. At all levels in the spinal cord and in the brain stem itself there are nerve fibers that cross from one side to the other. When I bump my left elbow, information goes to both halves of my brain, shunted there by any number of cross connections. On the other hand, primary responsibility for one side of the body resides in the cerebral hemisphere of the *opposite* side; in righthanded persons the left hemisphere is the dominant one. The dominant hemisphere is also responsible for language and speech. Why there should be a pronounced crisscross control arrangement between body and brain is unclear. The arrangement we see today could be a relic tracing back to the origin of the very first receptors when, perhaps, two receptors could be made to home in on an object by merely accelerating the motion of the side opposite the stimulated receptor. Simple mechanical robots are constructed in this manner.

The eyes and their associated optic nerves illustrate a crisscross arrangement of nerves connecting receptors and their corresponding (optic) centers of the brain. The accompanying diagram reveals that the exchange is more sophisticated than the very primitive scheme mentioned above. Each eye has a left and right field of vision. The individual nerve fibers leading from the right field of the left eye and from the left field of the right eye proceed to cerebral hemispheres without crossing to the opposite side of the brain. The fibers leading from the remaining two fields cross in the optic chiasma so that each lot proceeds to the cerebral hemisphere of the opposite side. As a result of this arrangement, the right side of the brain is informed of the left portion of the visual field while the left side is informed of the right portion.

The optic chiasma can be severed as shown in the diagram. Following this operation the left eye transmits messages only to the left half of the brain while the right eye transmits only to the right. Suppose now that a cat or a monkey with a severed optic chiasma is taught to perform a task with its left paw upon receiving a visual signal with its right eye. Can the animal do the same task with its right paw? Yes, provided that the great cerebral commissure is intact. Otherwise his right paw is unable to perform the task until the animal has been trained anew by signals directed at his left eye. The commissure carries experiences that have been recorded in one hemisphere and transcribes them in the other. By severing the great commissure, the animal gains what are essentially two separate and separately trainable brains.

When does the transcription of messages of one hemisphere take place in the other? In the cat it appears that transcription takes place immediately. Cutting the corpus callosum *after* a cat with a severed optic chiasma has learned through signals received by one eye to perform a task with one paw does not interfere with its ability to do it with the other. Man seems to operate differently. In man, experiences received by one hemisphere are recorded in that hemisphere only and are supplied to the other exclusively upon demand; in effect man doubles his filing space for memories. The monkey is intermediate to the cat and man; some experiences he transcribes immediately in both cerebral hemispheres whereas others are left in one until the other calls for them.

Persons who for medical reasons have had their great commissure severed (or, because of birth defects, have never had one) are unable to transcribe stored

VISUAL FIELD

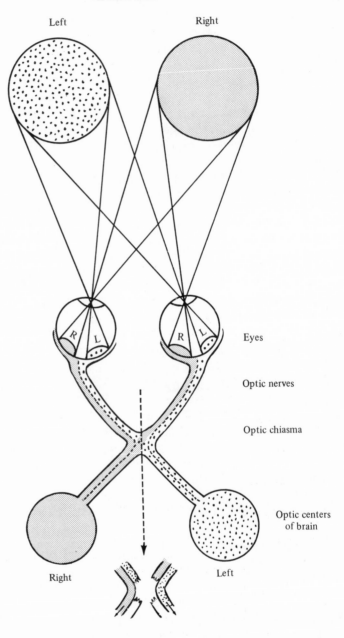

Left

Right

Eyes

Optic nerves

Optic chiasma

Optic centers
of brain

Right

Left

memories from one hemisphere to the other even on demand; the machinery simply is not there. In many respects these persons, like the experimental animals described above, behave as if they have two brains. By carefully aiming small beams of colored light, it is possible to shine red light, for example, onto the right field of the left eye and, simultaneously, blue light onto the left field of the right eye. The diagram shows that the red light will be perceived by the visual area of the left cerebral hemisphere and that the blue light by that of the right hemisphere. If the person is now shown a chart and is asked to identify the color of the light that was just shone into his eyes, his left hand will point to the color blue while his right hand will point to red. Or, instead of pointing with his right hand, the person may say, "Red." If he is right-handed, he will be unable to spontaneously identify blue by speech because language is the property of the dominant hemisphere, in this case, the left hemisphere.

A number of treatments (cooling, application of drugs, and electrical stimulation, for example) are known that temporarily inactivate the cerebral cortex without damaging the memories that are already recorded there. By the use of such treatments the mental capacity of a cat or rat may be enlarged. Different experiences can be recorded separately in the two cerebral hemispheres by inactivating the cortex of first one and then the other. It is known that experiences that have been recorded in one hemisphere of the rat's brain can be called upon to solve problems posed to the opposite hemisphere if the corpus callosum is intact. The solution to a problem is obtained very quickly by a rat using an "untrained" hemisphere if the answer has been prerecorded in the other half of his brain. By having two sets of experiences recorded, one in one hemisphere and the second in the other, both of which are needed to arrive at the solution to a problem, we obtain rats that are able to solve problems that otherwise would be extremely difficult. The two separate recordings are more effective than the sequential recording of the same two sets in the intact brain. An essential part of memorization is the assignment of a chronological order to the recorded experiences. Of two sequentially recorded experiences, the rat chooses the more recent one to provide a solution to a new problem; the earlier recording is effectively suppressed. In the case of a rat, one of whose hemispheres was inactivated for the first recording and the second for the next, the separately recorded experiences seem to carry no time label; consequently, the rat draws on both and therefore makes the most use of the information in his possession.

In this essay I have given a simplified account of an extremely complicated machine, the brain. Before concluding these remarks I must point out that there are many remote interconnections between the two cerebral hemispheres in addition to the great cerebral commissure. Some of these crossconnections involve components outside the body. The person who could point to the color blue but was unable to speak the word "blue" would speak it, of course, as soon as his eyes saw where his left hand was pointing. Or, if given two choices for a word that was unknown to the right hemisphere, a person can guess. Having heard (by way of its ear) a wrong guess, the opposite hemisphere can correct it immediately while apologizing, "Whatever made me say that; I know the answer perfectly well."

I should also emphasize that, following damage to the brain, new circuits are set up and new routes for nerve impulses are established so that in time the body functions quite normally. As one author has described it, the dominant and subservient portions of the brain are established as a matter of convenience so that orders will be issued and carried out in a methodical manner.* A dominant portion and a subservient portion of the brain correspond to a company commander and his orderly room clerk. Following a catastrophe that left the clerk as the senior company officer, he might do a commendable job in carrying on as company commander; conversely, the commanding officer in an emergency might be able to carry out the duties of his clerk. Similarly, portions of the brain that traditionally have been assigned certain routine chores to carry out can, in an emergency, take over and carry out other, more complicated chores in a thoroughly adequate manner.

4. The primitive brain

"One fine day a species woke up with a well-developed olfactory apparatus. It could smell its prey." Those words represent one psysiologist's view of the acquisition of new organs in the course of evolution, but they hardly represent the view of an evolutionary biologist, nor are they likely to be a reasonable description of events as they actually occurred.

Evolution is a process involving the slow modification of existing structures, not one based on the miraculous appearance of novel organs or organ systems. A species never wakes up with something new. Most species consist of individuals scattered far and wide over large geographic areas. These individuals occur in clusters known as local populations. Local populations live in discrete areas where conditions for the species' mode of life are adequately met. Mates of both sexes are encountered within these small populations. Offspring, too, are born to these populations. The rabbits in the small park behind my house find their mates within this park or, more rarely, encounter them within a block or

*John Grayson, *Nerves, Brain, and Man* (New York: Taplinger Publishing Co., Inc., 1960).

two; they almost never meet individual rabbits that live on the Cornell campus or on the more remote west shore of Lake Cayuga. So how could all rabbits (or any other species) "wake up" with a brand new trait shared in common by all individuals or hundreds or thousands of spatially isolated, local populations?

The nervous system, like all other organ systems, has evolved gradually with one advantageous alteration added here, another there — each spreading from its geographic point of origin as the responsible genes diffused, carried by migrant individuals, from one local population to another. Sudden spurts in evolution may occur when modifications of independent origin meet during their separate passage through a species and present the fortunate members of some local population with evolutionary possibilities that never existed within the species before. Irregularities in rates of evolution, however, do not contradict the claim that evolution is a process based on the gradual modification of existing structures.

The brain of very early animals was a simple swelling of the anterior end of the central nervous system. The typical mammalian brain of today developed slowly over millions of years from the much more primitive reptilian-style brain of mammalian ancestors. The earlier, simpler organ was not an incompetent brain, however, by any means. The brain of a salamander, for example, may be primitive in contrast to that of a cat, but in contrast to that of an earthworm or insect it represents sheer genius. A salamander stalking burrowing aquatic worms in an aquarium is a picture of stealth and cunning. The brain that was responsible for the survival of the remote ancestors of man and other mammals is still a functional part of the modern mammalian brain. It does not function in all its original roles, to be sure, but it does fulfill roles that are vital nevertheless.

The thalamus is one of the primitive centers of the brain. It was, in fact, the highest center of the brain at one time — a center that played (to the best of its ability) the role now played by the cerebral cortex. It was at that time the chief executive of the nervous system. Today it is the administrative assistant to the chief executive — no mean position as those who attempt to reach the president can attest. The thalamus monitors all incoming messages arriving at the brain by way of the spinal cord. It permits certain messages to continue to the cerebrum; other messages are shunted elsewhere for appropriate action. On its own initiative, the thalamus may dispatch down the spinal cord impulses that inhibit the very transmission of certain messages; pain, for example, can be completely blocked in times of stress. Not only is the thalamus responsible for deciding which messages will proceed on to the cerebral cortex, it is also responsible for arousing the cortex so that it will be able to receive and act on them.

The hypothalamus is also a portion of the primitive brain. It serves as the master regulator of the body. It fills this role through its interactions with the pituitary body and the endocrine system. Functions under the control of the hypothalamus are known rather crudely as "visceral functions"; hunger, thirst, body temperature, water balance, blood pressure, sexual behavior, pleasure, hostility, and pain form a partial catalogue of items under its control.

The hypothalamus has been the subject of some of the most spectacular experiments in neurobiology. Very thin, insulated wire probes can be inserted into the hypothalamus of living animals. A tiny electrical current passing through the probe stimulates the few cells lying immediately at its tip. The stimulated region is so small that regions less than 1/2 millimeter from one another may be stimulated independently. The effects are nearly unbelievable! A cat can be put into a rage – back arched, tail puffed, claws out – by the stimulation of one small region of its hypothalamus. A nearby region may give rise to terror. Still another may cause a cat to exhibit all the signs of extreme pleasure. One region when stimulated will cause a cat that has just finished eating to gulp down additional food as if it were starving. Another region inhibits an animal's appetite so that it will starve without eating a single bite. The dualities of function of the regions of the hypothalamus are characteristic of that region of the brain. One small area when destroyed will put an animal into a deep slumber from which it can scarcely be aroused. Another will keep the animal awake until it finally dies of lack of sleep. A very small injection of salt solution in the hypothalamus can cause a cat to drink prodigious quantities of water.

The functions described above are visceral functions, not only in the sense that they represent gut reactions but also in the sense that they are basic functions of all higher animals, certainly of animals with backbones. They are functions so necessary for the continuation of life that they cannot be risked in the hands of conscious control, except for short periods when certain actions can be postponed. The cerebral cortex exercises this inhibitory role that is manifest in the table manners, self-control, and other niceties of civilized life.

Perhaps the most important aspect of the thalamus, at least in respect to the remaining essays of this section, is the extreme pleasure its stimulation (at the proper site, of course) can give an individual. Experiments have been devised in which rats with electrodes implanted permanently in the pleasure centers are allowed to control their own electrical stimulation. Rats wired for auto-stimulation in this way have been known to give themselves 100 or more shocks per minute continuously for hours. If the procedure is altered so that they must press a switch several times in order to receive a single shock, they work frantically in order to receive the shocks as often as possible. If they are placed in an apparatus that gives them painful electrical shocks as they try to get to the pleasure control switch, they endure the painful shocks rather than forgo their pleasure. They voluntarily endure more painful shocks for this purpose than they would endure in order to obtain food even if they were famished.

Man, too, has his pleasure centers. One system includes much of the primitive brain: the more ancient region of the cerebral cortex and the thalamus – what was once the olfactory region, in fact. Stimulation of this system produces an immensely pleasurable feeling.

Beneath the conscious thought of the cerebral cortex, therefore, are the basic or primitive feelings that set the general tone for man's activities, that determine whether he acts as he does against a backdrop of rage, or of hunger, or of pleasure. The cortex may decree – despite the backdrop – that an individual smile and say, "No, thank you, sir." But the backdrop remains unchanged despite a veneer consisting of a small, thin mat of nerve cells that cover man's brain. Background emotions or background feelings are set by what eons ago were the ruling centers of the nervous system. Today these centers are relegated to the subservient roles of clerks and assistants. This demotion in rank does not destroy the importance of these ancient centers. In the nervous system, as in government and business, a large portion of all power necessarily rests with lower administrative echelons.

5. The individual nerve cell: to fire or not to fire

The central nervous system must analyze incoming messages arriving from a multitude of sensing devices, interpret these messages in the light of past experiences, decide what action is called for, and – in carrying out this decision – activate a muscle or other appropriate end organ.

How are these decisions reached? Appropriately enough, by a process closely resembling an election under which the majority rules. Imagine a small boy who is tempted by a piece of cake that has been left on the kitchen table where he can see it. A number of impulses urge him to take it; he is hungry, cake is good, and no one is watching, for example. On the other hand, there are contrary impulses; the cake is not his, the last time he took a piece he was spanked, and, even though he is hungry now, the cake will ruin his appetite for supper. Eventually, his hand will reach out and take the cake or it won't; if not, his legs will carry him out of the kitchen to a new confrontation and still another of life's decisions.

What was described above on a macroscopic scale happens continuously on a microscopic scale at millions of separate sites within the central nervous system. A nerve cell capable of activating a remote muscle fiber (see diagram on page 280) can be used as the central figure (I shall refer to this cell as the "central cell" subsequently) in place of the small boy of the earlier example. The long axon leading from the body of the nerve cell carries the electrical impulse that activates the muscle fiber whenever the body of the cell starts the impulse on its way, whenever the body of the cell "fires." The cell fires, however, only if

+ Excitatory
- Inhibitory

Muscle

the messages it receives from a number of other nerve cells amount to an order to fire; otherwise it remains quiescent. Among the second echelon of nerve cells that impinge on the central cell there are some that when activated tell the central cell, "Fire!" Others say, "Don't fire!" The central cell of our example receives these impulses, adds them, ignores those second echelon cells that refrain from voting, and then fires or fails to fire depending upon the summation of incoming messages. And, of course, the cells that I have called "second echelon" fire or fail to fire according to the instructions they have received from still more remote echelons.

In one part of the brain, a part that is important in governing our interpretation of and attitude toward the outside world, there is evidence of the existence of two transmitter substances (substances that effect the connection between the central cell and its second echelon cells). These two substances work in opposite ways. One of them, *norepinephrine*, excites nerve cells whereas the other, *serotonin*, calms them. These substances are made in separate nerve cells; while awaiting use they are stored in granules at the ends of the long axons of second echelon cells (serotonin in some; norepinephrine in others) next to the cell body of the central cell. When serotonin is released from the terminus of an axon and moves to appropriate receptor sites on a nearby cell body, the releasor axon votes, "Don't fire!" By witholding serotonin (that is, by not firing itself), the second echelon cell refrains from voting "Don't fire!" – the closest that it can come to voting "Fire!" On the other hand, when norepinephrine is released from second echelon cells and moves to its receptors on the central cell of our example, they vote "Fire!" When this substance is withheld, the nerve cell refrains from voting (the cell's equivalent of a "Don't fire!" vote).

The central cell of this example can be involved in any one of a number of possible situations. All second echelon cells can be quiet and not firing, in which case the central cell is quiet too. One or more serotonin-secreting cells can be firing; once more the central cell is quiet. One or more of the norepinephrine-secreting cells can be firing, in which case the central cell also fires. One or more of both the serotonin and norepinephrine-secreting cells can be firing; in this case the central cell (like the small boy in the kitchen) adds up the different impulses and does what the majority orders – either it fires or it does not.

Knowledge about the democratic process by which nerve cells operate has come from the experimental treatment of persons suffering from nervous disorders and, conversely, advances in the chemical treatment of these persons has come from a fuller understanding of the workings of the nervous system. Agitated persons are persons whose nerves are firing too often; they suffer from too much nervous activity. These persons respond to tranquilizers. A number of chemicals (consult the accompanying diagrams) serve to calm overly agitated persons. One, chlorpromazine, blocks the norepinephrine receptors. Another, reserpine, inhibits the loading of norepinephrine granules in the terminus of the axon and, as a consequence, increases the rate at which this chemical is destroyed by an ever-present enzyme of the nerve cell. Lithium salts, long known to have tranquilizing effects, shunt norepinephrine from the pathway leading to storage granules to one leading to enzymatic destruction.

Persons who are pathologically lethargic respond to treatment by chemical stimulants. Amphetamine stimulates the release of norepinephrine and, hence, the tendency of nerve cells to fire. Iproniazid, another chemical stimulant, inhibits the enzyme that normally destroys excess norepinephrine and, as a consequence, increases the norepinephrine level in axons and the frequency with which other cells are stimulated to fire. Imipramine, also a stimulant, retards the

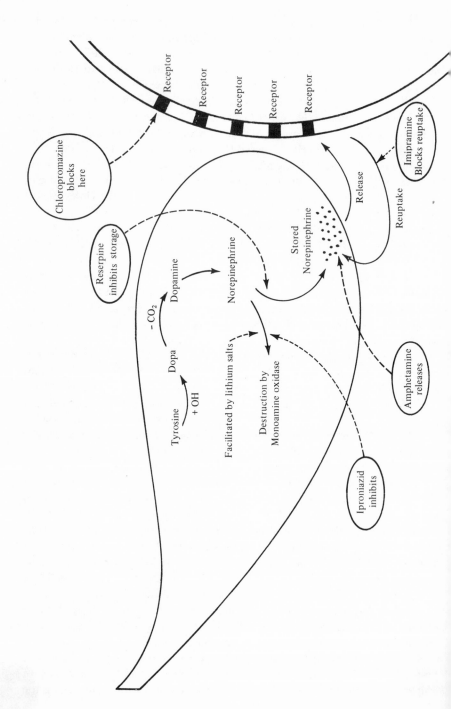

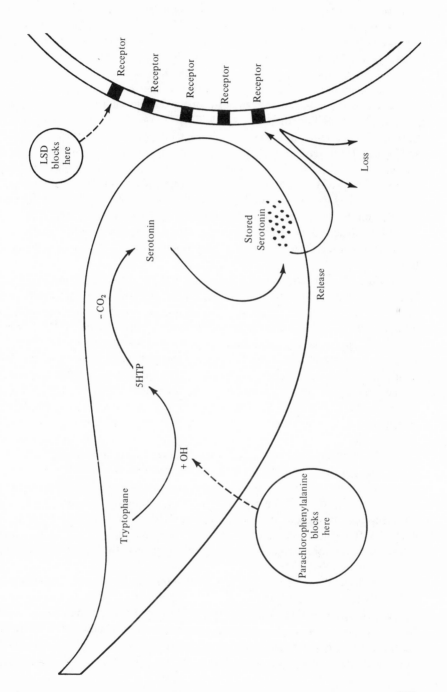

loss of norepinephrine from its receptors on the central cell and, in effect, biases each vote toward "Fire!" by stuffing the ballot box. Still another stimulant, parachlorophenylalanine, inhibits the formation of serotonin and, hence, inhibits the effectiveness of those nerve cells whose role is a calming one. Parachlorophenylalanine in this way inhibits negative votes in the central cell's role call and, as a result, makes it easier to sway the vote toward "Fire!"

The tranquilizers and stimulants that we have described here have been discussed as if they were used merely to correct disorders in the nervous system. Everyone is aware that they are also used by "normal" persons to manipulate their nervous systems into generally pleasant or tolerable sensations. Agitated and upset? Take tranquilizers and stop those nerves from firing. Lethargic and exhausted? A stimulant might pick you up. As long as these drugs have no lasting effect (wreak no permanent havoc within the nervous system), their use in an emergency can be tolerated. This is the basis for their use on a physician's prescription. Some of these drugs, however, if used for sufficient time, lead to a true physiological dependence; the body then requires the drug in order to continue operating. The narcotic drugs, morphine and heroin, are notorious in this respect.

Aside from the physiological dependence or true addiction, however, there exists a psychological dependence that grows with the use of mind-affecting drugs. In essence, these drugs alter the manner in which the nervous system interprets the outside world, in which it draws upon past experiences for making judgments; they may change the very pathways by which messages are relayed from place to place within the central nervous system. Even allowing for the contrasting interpretations that different persons will normally devise in respect to the same set of input messages, we must admit that the chronic use of drugs isolates the user from the external world. Under the influence of drugs he no longer responds to the world as it is, but as it seems to be or even as he might wish it to be.

The popular (and very dangerous) hallucinogenic drug, LSD, exerts its effect through the system of nerve activation described in this essay. It appears that LSD attaches to the serotonin receptors and, consequently, inhibits negative ("Don't fire!") votes. In this way, LSD acts as a stimulant. To the extent that this drug acts throughout the central nervous system, it can set up nerve cells so that they fire at unexpected times. Here is a plausible explanation for the visual patterns arising from sound and touch as well as for the other hallucinations that are induced by LSD.

My account of nerve action is a gross simplification of a formidable topic; very little about the nervous system is simple. Nevertheless, the account I have given is probably not terribly wrong even though the details may in the final analysis prove inaccurate. It is comforting, at any rate, to know that underlying nerve action there is a binary system — serotonin for calming and norepinephrine for stimulating — in at least one region of the brain. The binary system is a basic one with roots in evolutionary biology (the present genetic code appears to have evolved from an earlier binary one), in nerve impulses ("Fire!" versus "Don't fire!"), in primitive languages (bass versus alto beats of native drums, large versus small puffs in smoke signals), and in modern communication (dot versus dash in the Morse code, off- versus on- switches in high-speed computers). The details may well be altered by additional data, but the general outline that has emerged so far is attractive in its fundamental simplicity.

On the Inheritance
of Behavior

Several geneticists were dining recently in a Chinese restaurant on the West Coast. As the conversation drifted aimlessly from topic to topic, it touched on the contrast between Chinese and French cuisines. The heavy creamy sauces of the one, for example, are entirely missing in the other. The fancy cuisines do not differ, however, in an economical sense, for the ingredients used by chefs in the two cultures are about equally expensive. Because the diners were geneticists, the conversation also turned to an interesting genetic difference between Oriental and Western (European) populations. Adult Orientals generally lack an enzyme that is necessary for the metabolism of milk sugar; adult whites, on the other hand, generally have this enzyme. The Chinese cuisine lacks milk-based sauces; if adult Orientals were to eat such food, many would become seriously ill. Similarly, "Chinese restaurant syndrome" — an allergic reaction involving dizziness, nausea, and headache — appears to be a Western reaction to monosodium glutamate, a substance often used overgenerously in Oriental cooking in order to accentuate the subtle flavors of many ingredients.

To the best of my knowledge no one knows whether the absence of the enzyme needed for metabolizing milk sugar preceded or followed the origins of Chinese dietary habits. Babies of all races possess the enzyme in question because they digest mother's milk perfectly well; certain populations of people lose the enzyme after weaning. Here, therefore, is an adult food preference — one of the many facets of human behavior — that is involved with, if not actually based upon, a genetic difference between peoples. Not all food preferences need have corresponding sorts of genetic bases, of course. I, for example, have until now failed to avail myself of the chocolate-covered ants on sale at the local supermarket; I have refrained not because of enzyme deficiencies but because of prejudiced eating habits.

The distinction between those differences among men (or among individuals of any species) that have a genetic basis and those that are the result of environment or culture is not always an easy one to make. The problem has been encountered before in earlier essays. To grasp it and understand it, we must understand the meanings of "necessary" and "sufficient." The genetic material — the blueprint for life — that each of us obtains from our parents is *necessary* for growth and development. Similarly, nutrients and a place within which to grow and develop are also *necessary*. Neither genes nor nutrients are *sufficient* for growth and development; every individual needs both and, hence, reflects the influences of both.

In recent years the nature-nurture controversies of old have been revived. The controversy over the IQ of persons of different racial origins, for example, has erupted anew with the claim of Professor Arthur R. Jensen that the lower average IQ test scores of Negroes have a substantial genetic basis.* The appearance of several books dealing with the aggressive nature of man has evoked counterclaims that man's aggression has a cultural, not a biological, basis. Many psychologists believe the same to be true in respect to intelligence. In this essay I want to touch on a number of behavioral traits of lower organisms that can be clearly shown to have their roots solidly in the genotype. Man's much more complex behavior will reenter the discussion in the concluding paragraphs.

A great many organisms display unbelievably complex and intricate stereotyped behavior. Predatory wasps are an example. Females of these wasps prey upon other insects in order to provide food for their young. Females of each of a number of wasp species have a single prey species; in some instances the wasps are further restricted to capturing prey individuals of but one sex. With no training, these wasps seek out members of the single species on which they prey. When a victim has been found, it is very often permanently paralyzed (not killed) by an extremely accurately placed sting. Identifying its prey, paralyzing it, preparing the burrow in which it is placed, laying the egg on the prey, and closing up the burrow represents a series of acts performed by each female wasp as if she had spent long periods in intensive training. Each act, however, is part of an instinctive sequence that is run through much as a complex dancing doll performs endlessly on a Swiss music box.

Convincing proof that a pattern of behavior has a genetic basis requires that individuals with divergent behavior be mated and that the inheritance of these differences be noted in several generations of progeny. Evidence of this sort has been obtained for behavior differences in both honey bees and fruit flies. Rather complete studies have also been made with different breeds of dogs.

Flies with contrasting modes of behavior can be easily obtained in the laboratory by artificial selection. In one strain, for example, parents may always be chosen from flies that run from one culture bottle to another when they are given an opportunity to do so. In a second, contrasting strain, parents may be chosen from those flies that fail to leave the original bottle. The result within surprisingly few generations will be two strains of flies; one, nervous and active and the other, sedentary. Artificial selection of this sort has succeeded in producing house flies so lethargic that they permit themselves to be touched with the finger even when moving freely on a desk top.

---

*Arthur R. Jensen, "How Much Can We Boost IQ and Scholastic Achievement?" *Harvard Education Review*, 39 (1969), 1-123. See also the essay entitled "Race and IQ: A Critique of Existing Data" in Vol. II.

Dr. Jerry Hirsch, a behavioral geneticist, by means of artificial selection obtained strains of fruit flies that differed in their response to gravity. Individuals of one strain generally ran upward in passing across a vertical maze whereas those of the other strain tended to move down and across. By the use of special strains of flies that are kept in many laboratories for just such complex genetic analyses, Hirsch was able to mate and remate his flies in a systematic fashion so that new strains were obtained, each of which carried only one or the other of the chromosomes from a given selected strain. Individuals of these "reconstructed" strains behaved as would be expected if the modified behavior of the original selected populations had a genetic basis. Each of the individual chromosomes contributed a share (not necessarily equal shares) to the tendency of flies to run up the maze, or down.

Genes located on all chromosomes influenced the behavior of Hirsch's flies in the maze. In honey bees striking behavior has been observed that, upon analysis, proved to be the result of but two genes. The two strains of bees in this case differed in their response to a disease of bee hives called "foul brood." Larval bees infected by the disease die in the cells of the honeycomb within which they are developing. Worker bees in a hygienic hive uncap the cells containing dead larvae and dispose of the corpses; nonhygienic bees do not perform these cleaning duties. Genetic analyses involving crosses between hygienic and nonhygienic bee strains revealed that the two steps — uncapping the cell and removing the dead larvae — are inherited separately and that each is governed by a single gene. Thus, in addition to hygienic and nonhygienic bees, there are those that uncap cells containing dead larvae but do not remove the corpses. On the other hand, there are those bees that will remove dead larvae but will not uncap cells to get to them; in order to identify these bees the investigator must uncap the cells for them by hand.

That behavior has a genetic basis is surely not surprising. It has as its physical basis the nervous system, a system whose development is under genetic control in the same sense that any other organ system of the body is under genetic control. The nervous system is not immune to variation and, consequently, genetic differences among individual persons can result in behavioral differences. The evolution of the mammalian brain is good evidence for the genetic control of brain structures.

During the early 1900's, shortly after the rediscovery of Mendel's work on the inheritance of various characteristics in garden peas, geneticists struggled to show that many organisms other than peas exhibited the newly found "Mendelian" inheritance. The list of organisms studied in those years was tremendous: flowers of all sorts, vegetables, domestic animals, fowl, and insects. Cats, dogs, sheep, geese, corn, Jimson weeds, delphiniums, and fruit flies were subjected to investigation at one laboratory on Long Island. The inheritance of certain traits in man was also investigated at this time. In retrospect, it seems that many of the conclusions of these early studies were based more on enthusiasm than on substantial observations. Thus, one study concerned a certain woman whose father was a notorious vagabond and who had two sons – one studious, the other wild. The woman herself was a good wife and mother. Hence, a recessive gene for nomadism was said to reside on the human X-chromosome; the passage of a trait from grandfather to one-half of his daughter's sons is characteristic of X-chromosome inheritance. In another study, supported by the Navy Department in an effort to learn the source of the most promising recruits, a hypothetical gene for thalassophilia, for the love of the sea, was "found" to be very common among persons residing near the seacoast. Obviously, conclusions of these sorts were based on the most simplistic models; such conclusions could, in fact, be seriously entertained by only the most uncritical minds.

Do the above remarks tell us that the human mind and human behavior is independent of one's genotype and of the body this genotype helps construct? Not at all. The importance of the physical structure is revealed by experiments on the electrical stimulation of the brain. That a rage can be artificially induced, for example, by a small electrical current in the absence of any other outside stimulus tells us of the importance of the physical structures of the brain in controlling our emotions. Rage – no rage – pain – pleasure – rage again – all can be effected by shunting electrical currents to a number of small electrodes inserted here and there in the brain. Perfect and extremely vivid recall of events long past may follow the stimulation of the temporal lobes of the brain with a small electrical needle. Once more we see the importance of the physical structure of the brain itself. Like all of the body's physical structures, of course, its construction is subject to genetic control.

The human mind appears to be largely independent of the individual's genotype, it seems to me, because of the redundancy that is built into it, especially into the cerebral cortex. If the cerebral cortex were spread out as a flat sheet, it would form a square 18 inches by 18 inches. The cells, 10 billion of them, in this sheet are interconnected to receive nerve impulses, shunt them about for evaluation, and pass on appropriate directions to those muscles of the body that should be made to respond in a suitable manner. These rational processes can inhibit irrational ones whose reflexive origins lie deeper in the brain; such inhibitions are the basis of civilized behavior.

After suffering a disabling stroke, many persons eventually regain the use of their paralyzed limbs. Why they should be able to do so is not well understood. One possibility is that certain cells in the cortex are temporarily disabled, but not killed, as a result of the stroke. As normal circulation is restored in the brain these cells once more resume their routine functions and, as a result, the individual slowly returns to normal. Alternatively, neurons whose original functions were not overwhelmingly important may have been withdrawn from their original duties and, through retooling, pressed into other services. If the latter possibility should prove to be even partially true, it would reveal why many persons believe that the mind is formed exclusively by environmental events impinging upon it from outside the body. These persons (environmentalists) would be correct *in part*. They would be correct to the extent that the full capacity of the mind exceeds the demands that most of us place on it during any one segment of our lives. It is this excess capacity that lends itself to manipulation at the hands of the behavioral psychologist.

On Kidding Ourselves

Scattered through our lives are moments when circumstances thrust upon us a need to make decisions that for practical purposes are irreversible. We choose to attend a certain college; we accept a particular position; we marry; or we buy a house. In each case we must live with the decision for an extended period of time. How does a person adjust mentally to the choice he has made? Or does he adjust? This essay will be concerned with these questions.* The title reveals that adjustments in attitude favorable to the choice are made. To whatever extent these changes in attitude are not dependent upon newly acquired information but represent instead an accommodation to an unalterable circumstance, they illustrate man's tendency to kid himself.

Inducing a person to lie creates as stark a contrast between two alternatives as we can imagine: a person who lies has at a given moment a choice, and during that moment he chooses to say that which he knows is untrue. How does he cope mentally with this conflict of information? It seems that much depends upon the original inducement to lie. If a substantial reward is received in payment for making a false statement, the person tends to look upon his statement as false. If little or no reward is received, however, the person tends to convince himself that what he said was not a lie but, on the contrary, was the truth.

Experiments that furnish information supporting the claim that persons may come to believe their own lies have been made with students. In one instance, those who had finished a dull chore were asked to tell their incoming replacements that the job had been great fun. For "helping" the project in its recruitment of new subjects by perpetuating a myth of "fun," the students were paid either $1 or $20. In a subsequent interview dealing with their attitudes in respect to the chore they had performed, those who had been paid $20 admitted that it had been exceedingly dull. Those who had been paid $1 claimed, on the contrary, that it had been a good deal of fun and rather stimulating. Only by convincing themselves of this fact, apparently, could they come to live with the knowledge that they had passed on false information.

Another experiment of the same sort consisted of enlisting students to write essays advocating opinions that were contrary to their own beliefs. For performing this task, the students were paid either $0.50, $1, $5, or $10. Later these students were tested on their beliefs in respect to the essays they had written. Those who had been paid $10 were even more strongly opposed than

*Material relevant to this essay appeared in an article by Leon Festinger entitled "Cognitive Dissonance," *Scientific American*, October, 1962, 93-102.

they had been before, but those who had been paid only fifty cents now professed to believe much of what they had written. Students who had received $5 and $1 had altered their views so that they also tended to believe what they had written. The results of the tests on the four groups formed a regular pattern extending in an orderly manner between the $10 and 50¢ extremes.

In these experiments we see an explanation for some of the inconsistency of man's actions noted by Montaigne. In a sense, a person alters his beliefs to maintain his self-respect. If the reward for performing an incongruous act is great enough, however, self-respect remains uninvolved. Presumably pride of a sort predominates: "Of course I lied; look what they paid me!" Self-respect, in this case, is assuaged by the receipt of a sizable reward. The ebb and flow of rewards granted at various moments in life and the need to preserve his self-respect when false actions are performed for little or no reward lead to the patchwork appearance of man and his actions that was described by Montaigne.

The evidence I have cited here is not unrelated to the restlessness of young people described by Paul Goodman. There are, of course, many reasons for today's unrest. Over us all hangs the threat of a devastating nuclear war. In addition, our earth is being exploited by vast numbers of energy-hungry human beings. These matters occupy other sections of this book. That much of the terribly wasteful hustle and bustle going on about us concerns trivial matters possessing no rational justification is apparent to our youth now — just as it was years ago to Thoreau. Goodman's *Growing Up Absurd* condemns modern society for suffocating its young.

If society and its aims will not change, this essay must end on a bleak note. Here are young people who, looking in from the outside, recognize the absurdity of the modern technological system. Once they enter it, they must embrace it to preserve their self-respect. The extent to which they will embrace it is inversely proportional to their status within it. Those at the bottom who receive the least reward, like the students who were paid 50¢ to write a lie, must adjust mentally — must live a lie — in order to justify their actions to themselves, to their wives, and to their children. Others, placed higher up, can afford to remain cynical.

If dissatisfaction with the aims of society were to become widespread, however, so that these aims were changed, the young would have a world into which they might enter with pride. They have risen to meet various situations on a number of recent "children's crusades"; they will probably rise to meet new occasions. Eventually society will change. It will do so, however, only through pressure, and pressure can be applied only by those active participants in society who are in possession of their senses. No pressure is applied by those who isolate themselves. No pressure is exerted by those who withdraw into a dreamland of drugs. Much of Haldane's long life was spent applying pressure to governments in order to evoke social reforms — a fact that brings us to his claim once more: "The alterations of my consciousness due to . . . drugs were trivial compared with those produced in the course of my work."

On Escaping Reality

It is eight-thirty in the evening as I begin to write. Nearly four hours ago I finished scanning *The College Drug Scene* and *Science, Conflict, and Society*, two books that recently arrived in the mail. What have I done during the past four hours? I enjoyed a martini or two, had dinner, and relaxed with the family. Alcohol is the most widely used of all drugs. In this world there are more persons who are addicted to alcohol than to all other drugs combined. Despite the danger of addiction, nearly every evening I treat myself to a predinner drink with all of the ceremony that goes with the mixing of an extradry martini.

The cocktail hour is a moment of escape from the daily routine of life. For a short time the worries over deadlines, reports, orders, lectures, and reading assignments recede, and as a result the world appears less formidable and a bit more tolerable. Now is the time to reminisce with the children. We talk of the dachshund braving the waves near Barcelona in order to retrieve a rubber donut for his master. We recall an evening on the Acropolis when we listened to a rehearsal of Aida being held far below in the amphitheatre. We talk of the obliging porpoise that came to the window at the San Francisco zoo in an effort to get his head rubbed. Escape – pure and simple – but also fun. The cocktail hour is a time for the family to dream again of yesterday's pleasures and to scheme for those of tomorrow.

Dinner, the news, one of several preposterous TV shows, and then back to work. Here is the miracle – the genuine miracle – of alcohol. *Its euphoric effect is temporary*. By the time that the newscasters have described the latest oil slick, the derailments, and have given the latest casualty figures for Southeast Asia, the real world is back once more in all too sharp focus. The spectacular escapades of the latest TV hero offer me a last glimpse of unreality before I return to my study.

There are many paths by which man can escape from reality. Indeed, some persons neither want nor ask for escape. Either these persons have a tremendous enjoyment for life as it is (or seems to be) or, through unwavering inner convictions, they believe in perpetually facing up to reality. Other persons have luncheon cocktails, dinner cocktails, after-dinner drinks, or any combination of the three. The lost weekend is still another form escape may take, and beyond the lost weekend there are the lost years – the years spent by the thousands of derelicts on the skid rows of the world's cities. Selfdestruction is the last escape offered by alcohol.

Already the discussion has brushed against a serious dilemma. Where do we stop? Who says what is right? The alcoholic on skid row, even though he may be escaping the realities of life, is regarded by most persons as a pitiable object. Why? Everything we see, or hear, or feel is seen, heard, or felt only in terms of electrical nerve impulses that flash through complex circuits of the brain. What then is wrong with setting the switches so that these impulses, no matter what their origins, bring something akin to pleasure?

As we saw in a previous essay, rats can be experimentally equipped with permanent electrodes in the pleasure centers of their brains. If these electrodes are wired to a lever that the rat can operate himself, he will remain there for hours at a time shocking himself as many as 5000 or 7000 times each hour. A rat will operate the switch to the utmost limit of his physical ability. In some manner the pleasure that the rat obtains by the electrical shocks is related to sex because self-stimulation is drastically reduced after castration, only to be resumed following the injection of testosterone, a male sex hormone.

Institutionalized persons have also been outfitted with electronic devices by which they can administer small electrical shocks to their brains. These devices have been developed for certain mental disorders where the patient can sense the approach of an otherwise convulsive seizure and prevent it by giving his brain an artificial, electrical stimulation. An automatic counter that was attached to one such device revealed that the patient (male) was repeatedly pressing only one of the three switches that operated different electrodes embedded in his head; when questioned he admitted that the button he preferred to press gave rise to strong sexual feelings resembling (but not culminating in) an orgasm.

Electrical shocks (small ones, of course) have an air of neatness about them. If electrodes are embedded in the proper site in the brain, a touch on the switch can give rise to an immensely pleasurable sensation. What could involve less mess or less bother? Of course, if a person loses control of himself, as does the rat, and does nothing but press the switch, then the original dilemma returns. The brain receives impulse after impulse that is interpreted as pleasure, but to those watching from the outside the individual begins to fall apart — physically and mentally. Pleasure, we might recall, is a sensation whose origins are in the primitive brain, not in the cerebral cortex, which is the center for rational thought. The nymphomaniac is a woman who is also dependent upon the repeated stimulation of the pleasure center of her brain (by coitus rather than by electrical shocks); despite the thoughtless and callous reactions of many people, the nymphomaniac is an ill person deserving sympathy and help.

Disease can affect the brain stem and distort the sensations it perceives. Kuru, a degenerative disease of the central nervous system, is such a disease. It affects the natives inhabiting a small region in the mountains of New Guinea. More than half of the deaths of these natives are caused by kuru. At one time kuru seemed to be a genetic abnormality, but more recently it has been blamed on a strange virus-like particle apparently transmitted from person to person, from the dead to the living, by cannibalism — by eating infected human flesh.

The early symptoms of kuru include an unsteady gait and tremors, both of which worsen over a period of several months until the patient is unable to walk or stand. Intelligence is not affected, but speech becomes slurred. Of special interest is the following description:* "The patients often display a marked emotionalism, with excessive hilarity, uproarious, foolish laughter on slight provocation, and slow relaxation of emotional facial expressions." Postmortem examinations (the disease is invariably fatal) show degeneration of nerve cells in the primitive regions of the brain, the regions where the pleasure centers are known to be located.

Sheep suffer from a disease, scrapie, that closely resembles kuru. Like kuru, scrapie is a neurological disease caused by a virus-like particle. The disease requires several months to run its course. And, again, the pleasure centers of the brain seem to be involved. Farm youngsters often have pet lambs that are spoiled by the scratching and affection they receive. These pet sheep exhibit their pleasure by arching their backs and by making characteristic nibbling motions with their lips. In a region where scrapie exists, a veterinarian is frequently at a loss in attempting to diagnose an isolated case; the suspected sheep may be either one suffering from scrapie or the family pet.

Alcohol in excess, nymphomania, and kuru – this is a shabby roster of pleasure-stimulating agents or procedures. None of them is particularly attractive when viewed by an outsider at any time. None of them, especially in the advanced or terminal stages, can be particularly pleasant to the individual involved. To this roster we must add hallucinogens and other drugs. Those drugs that lead to a physiological dependence certainly must be placed on the list because, in exchange for some short-lived albeit spectacular sensations, the individual unwittingly makes a lifelong commitment. He might as well arrange for contracting kuru. The early stages seem to be equally pleasant and, in many instances, the terminal stages may not differ very much.

Hallucinogens that do not lead to physiological dependence (addiction) or cause permanent physical harm to the brain or other tissues are not necessarily bad unless their use is abused. By "abuse" I mean using a drug with the notion that the sensations arising from its use should become the prevailing sensations of life. It makes no difference, in my opinion, whether the dependence on hallucinogens that leads to abuse is psychological or physiological; the continual, repeated use of a drug for maintaining a permanent sense of unreality is a pathetic practice. It produces, in effect, an imitation of the symptoms that are caused by certain brain tumors.

There is nothing mystical or profound about disarranging our nervous systems. The effect of LSD, for example, may be more complicated than that

*D.C. Gajdusek and V. Zigas, "Degenerative Disease of the Central Nervous System in New Guinea," *The New England Journal of Medicine*, 257 (1957), 974-978.

caused by a blow on the eye, but it is not fundamentally different. In the one case, pressure is interpreted as "light"; in the other, a great many senses are deranged simultaneously. I prefer, during most of my waking hours, to look at the world and see problems much as others do; this is what I call reality. If I see solutions for these problems that other persons have not seen, or if I see flaws in solutions that others have suggested, I contend that I have added to the total store of human knowledge. In short, I say life is worthwhile. If, on the other hand, I see an adult person who is sitting cross-legged on a stage and discussing visions that only he can see, or visions that I can see only by rewiring my nervous system and by short-circuiting the evaluation center of my brain, I am intrigued. I am intrigued, that is, until I realize that these visions are being promoted as a permanent way of life rather than as an occasional, brief experience. Then I think of the prognosis for kuru and turn away. A brief experience, a brief respite from reality, I can understand. Some persons may choose marihuana, but the break I understand best is the one called the cocktail hour. There are persons who cannot tolerate the illusions (and boring conversation) that cocktails create; obviously, these persons do not drink. That is their privilege. I do hope that they have retained an ability to dream however. Otherwise, where will they come by the flashes that help them out of the rut of everyday life? Without dreams, where will these persons encounter the bizarre relationships that break the constraints governing man's hackneyed view of his world?

wouldn't there
all of what be
have been —

he wound of a
copy and these
history them
history dony

Claire
dream and their
all your I
Front
how...

ultimately
the answer lies
with the image
in the view — me

screen
—

LSD no somebody's
dream into
the night
the pen becomes
my finger hear
shall dry
the murderer
slinks into
the face
— that will see —

Two pages of a record of trips kept by a college student now deceased.

On Obedience and Conformity

> I observed a mature and initially poised businessman enter the laboratory smiling and confident. Within 20 minutes he was reduced to a twitching, stuttering wreck, who was rapidly approaching a point of nervous collapse. He constantly pulled on his earlobe and twisted his hands. At one point he pushed his fist into his forehead and muttered: "Oh God, let's stop it." And yet he continued to respond to every word of the experimenter and obeyed to the end.*

Here is a vivid account of the grim fate of a normal neighborly-type businessman. What happened to him? Was he subjected to torture? Is this an example of brainwashing as practiced by a sinister foreign power? Not at all. It is a description of a man who thought that he was torturing someone else. If he became so terribly distraught, why didn't he stop? That's the point! Although he was under no compulsion to continue and although he was approaching a nervous collapse, he found himself unable to say, "Enough!" This businessman is not an aberrant freak. Two-thirds of the men who volunteered for the same task that he undertook were also unable to break off their role as torturers although each showed an equally severe emotional trauma.

The experiment that involved these men was carried out at Yale University. The subjects were men, 20 to 50 years of age, living in New Haven and surrounding communities. Each subject, together with a stooge, was brought to the laboratory and advised about the alleged purpose of the experiment in which they were to take part: an investigation of the role punishment plays in learning. One of each pair (subject and stooge) was to be the student and the other, the teacher. The student, strapped in an "electric chair," was to learn word-pairs that were to be read to him by the teacher, who sat out of sight before an impressive control panel in a separate room. The list of word-pairs was read by the teacher, who then reread the first word of each pair and asked the student which of four other words was the second member of the pair. An error on the part of the student at any point in the list called for an electric shock as punishment. Following an error the entire list was then read to the student again; the reading was followed once more by a multiple-choice test. Another error called for a still greater shock and for still another repetition of reading and testing.

The 30 switches on the electrical control board by which the shocks were to be administered were labeled from "15 volts" to "450 volts" in steps of 15

---

*Stanley Milgram, "Behavioral Study of Obedience," *Journal of Abnormal and Social Psychology*, 67 (1963), 371-378.

volts each. The switches were arranged in groups of four, above which were the following verbal descriptions: *SLIGHT SHOCK, MODERATE SHOCK, STRONG SHOCK, VERY STRONG SHOCK, INTENSE SHOCK, EXTREME INTENSE SHOCK, DANGER: SEVERE SHOCK,* and, above the last two switches, an ominous *XXX.* Needless to say, the subject was always designated the teacher in this experiment (both slips of paper presented to him for drawing lots bore the word "teacher"); the stooge was not subjected to any electrical shock although this fact was unknown to the teacher; and the real purpose of the experiment was to determine how severely one person (the teacher) would be willing to punish a seemingly poor learner.

Upon each reading of the list of word-pairs and the following test, the student was instructed to make one or the other of several errors; in short, he never made a perfect score. Each time his error called for an increased punishment. When the punishment reached *INTENSE SHOCK* (300 volts), the student pounded on the wall of the room in which he was seated. After "receiving" 315 volts, he no longer gave a response to any question asked of him. At this point the teacher generally glanced at the experimenter for guidance and was told to treat silence as an error, that is, to increase the punishment by raising the voltage another 15 volts and to reread the word list again. The only encouragement to proceed offered by the experimenter to the teacher (the subject of the test, remember) were these phrases: "Please continue"; "The experiment requires that you continue"; "It is absolutely essential that you continue"; or, as a last resort, "You have no other choice, you *must* go on."

Not one of the 40 subjects tested, ordinary citizens of the sort we encounter every day (of *our* sort, if you wish), stopped administering shocks before reaching (what each thought was) the 300-volt level. Only 5 of the 40 (1 in 8) stopped at 300 volts upon hearing the student pound on the wall. Another 11 could not bring themselves to continue subjecting the student to shocks of still higher and higher voltage after he had ceased responding; these voltages were those clearly (though fraudulently) marked *EXTREME INTENSE SHOCK* and *DANGER: SEVERE SHOCK.* The disturbing point, however, is that 26 of the 40 (2 in every 3) continued to the highest possible voltage (one marked *XXX*) and continued into this range at the merest prodding by a stranger in a white laboratory coat, an experimenter whom they had never seen before nor were ever likely to see again.

Those subjects who continued giving high-voltage shocks to their "students" did so only with severe emotional strain. Hidden observers watched them sweat, tremble, groan, bite their lips, and dig their nails into their flesh. Half of the obedient subjects, those that continued to the very highest voltages, laughed nervously each time they flipped a switch. Several of them succumbed to uncontrollable fits of laughter. At the end of the experiment, the subjects sighed audibly, mopped their brows, and fumbled for cigarettes. For each, the experience had been an obviously harrowing one but one they were unable to terminate voluntarily.

   The results of this study reveal the seeds for otherwise inexplicable events such as the massacre of civilian women and children at My Lai or the mass extermination of Jews in Nazi Germany during the 1940's. The notion of obedience is carried by each of us. Under most circumstances obedience is an advantageous trait. A great many of us owe our lives several times over to instinctively obeying shouts of "Look out!" or "Duck!" Such warnings have prevented us from being struck by cars, falling objects, baseballs, and other missiles. In national affairs, too, there are emergencies when people must act in concert to a much greater extent than otherwise. United action by large numbers of persons depends upon obedience. The remarkable (and disturbing) degree to which persons are guided by obedience emerges clearly, however, only when groups of citizens perform acts that literally lead to national shame or, as in the study just described, when single individuals reveal how trivial the authority must be that drives them to perform acts of which they themselves are ashamed.

   Matters of social pressures as well as of obedience are involved in concerted human behavior. A great many persons "hold" beliefs, for example, only if these correspond to those expressed by others around them. This attitude permeates to even minor physical differences between persons as if these, too, could be right or wrong. Very early in this century, Dr. A. F. Blakeslee noticed that flowers do not smell the same to different persons. Some persons are unable to detect any odor from a particular flower; still others may differ in saying that the odor is either sweet or pungent. Different types of flowers elicit a variety of reactions from the same persons; those who are unable to smell one type can smell another, and vice versa.

   These almost casual observations led Dr. Blakeslee to set up demonstration booths at scientific conferences and other meetings in order to determine the extent of this variation in human populations and, if families were present for testing, its mode of inheritance. The point that struck him as the strangest and perhaps most disturbing was the manner in which many of his test subjects would loiter near the booth and then approach to ask, "Was my answer right?" These persons were unable to understand that individuals normally differ one from the other and that there is no right or wrong in smelling or not smelling a flower. The presumption that there must be a right and wrong for each set of alternatives together with the desire to be counted among the "rights" caused many of Blakeslee's test subjects to become suspicious of and irritated by his explanations. These persons were thoroughly convinced that he was hiding something from them.

   Some startling facts have been revealed by experiments with groups of college students in which all members but one gave erroneous answers in judging the relative lengths of lines. Subjects who were nearly always correct in their judgments when tested alone made errors in nearly 40 per cent of all trials when tested in a group where the other members had been coached to give incorrect answers. Among the subjects there were some who were unaffected by the group's opposing views and others who *always* followed the opinion of the majority. The effect of the group was felt if it consisted of three or more persons. A conforming person may dare disagree with one or two others but will give in to higher numbers. The influence of the group could be largely undone if just one of its members other than the subject made correct decisions. This bolstering effect was maintained even if the newly found "crutch" left the room. If, however, he deserted the subject and joined the remainder of the group in their erroneous decisions, the effect on the subject was disastrous — the frequency of his errors shot up immediately.

What is the meaning of these diverse findings? They mean that in a society where decisions are supposedly made on the basis of many independent opinions, the opinions are in reality not independent. A great proportion of persons watch others to see what they say or do, and then speak or act in conformity. Such a pronounced need for conformity is worrisome enough when viewed by itself. When, however, we combine the urge for conformity with the nearly blind tendency of persons to obey instructions from "above," we can begin to appreciate the tightrope that a democratic society walks. We can now see why many citizens are gravely concerned by political pleas addressed to the "silent majority." We can also understand the dangers inherent in any form of censorship, in the wholesale imposition of an "official" view on news reports. The silent majority of Germans, after all, condoned their government's racial policies during World War II. A large segment of any population clings together not from a consideration of right or wrong but for conformity's sake alone. The majority of these same persons have within them a tendency to render blind obedience to those in authority; and so, the seeds of national tragedies exist in all societies. There, but for the grace of God, go we.

Epilogue

The three volumes I set out to write are now complete. As I leaf through the various essays, I am satisfied that here indeed is a biology "textbook" unlike any other that I know. As I reread some of what I have written, I feel decidedly uneasy because on many issues I have dared express a personal point of view. I have, in short, exposed myself to the student. It is not easy to flout the traditional notions that a scientist should remain aloof from worldly problems and that a textbook should be a detached and academic recital of numerous impersonal facts. To see the extent to which I have expressed opinions on peripheral matters is as disconcerting to me as it will be to others.

My confidence is partially restored when I recall the readers for which these volumes are intended — for the terminal college-level biology course for the nonbiology (and, probably, nonscience) major. This student in all likelihood has had a modern high-school biology course comparable to one of those prepared by the Biological Sciences Curriculum Study. Consequently, he has been exposed to traditional biological facts. My collection of essays is intended to warn this student of and prepare him for the confrontations he will encounter in society — confrontations that involve the growing numbers of persons, the growing energy demands, the spread of Western-style industrialization to other lands, the pollution of the environment, the destruction of the seas, and the depletion of natural resources. These confrontations will require that he be equipped with more than standard biological information so that he can withstand the "image makers" and their seductive accounts of "progress."

In many respects, the information gained by a student in the usual biology course is inadequate for his later needs. To a large extent this inadequacy can be traced to the professed need for evaluating each student's progress and for determining his class standing. Unfortunately, trivial matters can be evaluated much more readily than substantive ones. Because large classes lead to an increasing reliance on machine-scored multiple-choice examinations, trivia now threaten to dominate course material — material that appears designed to fit evaluation mechanisms as often as the reverse. In retrospect, I am satisfied that in writing these essays I have not catered to the quiz makers. I have not assembled a mere collection of convenient test items. It is unimportant whether or not the student agrees with either the positions I have taken in various essays or the reasons for which I have taken them. If he disagrees, however, let him assemble both data and arguments in support of an alternative position. And let his ability in this respect be both the source of his own personal satisfaction and the basis for his evaluation by others.

The struggle to bring society's problems under control so that man will have a decent and pleasant world in which to live will not be easy. Because science and technology have been largely responsible for today's environmental ills, many scientists and technologists have regarded calls for corrective measures as attacks against themselves as well as against human reason. In carrying out a multitude of activities on what is truly a circumscribed and finite earth, these persons for decades have employed arguments, calculations, and technical procedures suitable only for an open universe of limitless dimensions. It is true, of course, that scientists and technologists will eventually help solve the very same problems that they have created, but their help will be effective only if their decisions are subject to scrutiny (and veto, if necessary) by other *equally knowledgeable* persons.

At a time, for example, when overcrowding, overconstruction, and air pollution in southern California are national scandals, it strikes me as sheer madness to speak glowingly of converting Baja California into still another southern California through the use of nuclear-powered desalinization plants.* In addition to missing entirely the basic role of population growth in causing the overcrowding of presently inhabited areas, the proposal would for all time turn the Gulf of California and the waters of the Pacific coast into sterile bodies of warm, concentrated brine. "Externalities" such as the need to dispose of heat, salt, and radioactive wastes cannot be neglected in a finite world. Those who do not understand this simple point will never be able to devise lasting solutions for any of today's serious problems. On the contrary, the continually optimistic and expedient suggestions of these persons lend credibility to MacDermott's dismal view of technology.

The favorite phrases recited by those who insist on forging ahead on the disastrous course that industrialized societies have set for themselves are "You can't stop progress" and "You can't turn back the clock." Both claims are wrong. In order to preserve the earth on which we live, we can do what must be done. If the clock must be turned back, it will be. In 1855, a certain Commander Rodgers reported to the U.S. Secretary of the Navy from Japan: "These people seemed scarcely to know the use of firearms. It strikes an American . . . that ignorance of arms is an anomaly indicative of primitive innocence and Arcadian simplicity." Unknown to the commander, the Japanese islanders of whom he wrote had been the first Japanese to adopt firearms; by 1700, after a century-and-a-half of use, firearms had been abandoned, however, because their use had raised more problems that it had solved. What those Japanese could do in ridding themselves of firearms, we can also do in ridding ourselves of much of the useless, costly trappings of modern life. If eliminating wasteful industrial practices means stopping progress, if zero population growth means stopping progress, and if a stable rather than an expanding gross national product means stopping progress, then this book has been written to encourage a host of progress stoppers.

---

*Gale Young, "Dry Lands and Desalted Water," *Science*, 167 (1970), 339-343.

Appendix A

The following essays in volumes I and II discuss matters related to the sections of this volume:

Disease

Communication

Appendix B

Table of Contents, Volume I

On the Stability of Biological Systems

Reaching a Simple Decision

Brinkmanship: The Limits of Game Theory

Table of Contents: Volume II

SECTION ONE: GENETICS

Introduction

GROWING UP IN THE PHAGE GROUP, *James D. Watson*

THE DOUBLE HELIX, *James D. Watson*

DNA: Blueprint for Life: An Essay in Three Parts
1. *Self-replication as the basis for life*
2. *The control of physiological processes*
3. *Heredity and individual development*

The Central Dogma: The Preservation of Simplicity

Genetic Engineering: The Promise of Things to Come

The Impact of Genetic Disorders on Society

Genetics Under Stalin: The Suppression of a Science

SECTION TWO: EVOLUTION

Introduction

THE VOYAGE OF THE BEAGLE, *Charles Darwin*

On the Evolution of Insular Forms

Natural Selection: Directional and Stabilizing

Meeting Conflicting Demands

Instant Speciation

On Modern Genetics and Evolution

Index

An alphabetical listing (based on **KEY WORDS** of their titles) of the essays of this volume.